Artificial Intelligence: Foundations, Theory, and Algorithms

Series Editors

Barry O'Sullivan, Department of Computer Science, University College Cork, Cork, Ireland

Michael Wooldridge, Department of Computer Science, University of Oxford, Oxford, UK

Artificial Intelligence: Foundations, Theory and Algorithms fosters the dissemination of knowledge, technologies and methodologies that advance developments in artificial intelligence (AI) and its broad applications. It brings together the latest developments in all areas of this multidisciplinary topic, ranging from theories and algorithms to various important applications. The intended readership includes research students and researchers in computer science, computer engineering, electrical engineering, data science, and related areas seeking a convenient way to track the latest findings on the foundations, methodologies, and key applications of artificial intelligence.

This series provides a publication and communication platform for all AI topics, including but not limited to:

- Knowledge representation
- Automated reasoning and inference
- Reasoning under uncertainty
- Planning, scheduling, and problem solving
- Cognition and AI
- Search
- Diagnosis
- Constraint processing
- Multi-agent systems
- Game theory in AI
- Machine learning
- Deep learning
- Reinforcement learning
- Data mining
- Natural language processing
- Computer vision
- Human interfaces
- Intelligent robotics
- Explanation generation
- Ethics in AI
- Fairness, accountability, and transparency in AI

This series includes monographs, introductory and advanced textbooks, state-of-the-art collections, and handbooks. Furthermore, it supports Open Access publication mode.

Nardine Osman
Editor

Electronic Institutions

Applications to uHelp, WeCurate
and PeerLearn

 Springer

Editor
Nardine Osman
Artificial Intelligence Research Institute
Spanish National Research Council
Bellaterra, Spain

ISSN 2365-3051   ISSN 2365-306X   (electronic)
Artificial Intelligence: Foundations, Theory, and Algorithms
ISBN 978-3-319-65604-5   ISBN 978-3-319-65605-2   (eBook)
https://doi.org/10.1007/978-3-319-65605-2

This Springer imprint is published by the registered company Springer Nature Switzerland AG
The registered company address is: Gewerbestrasse 11, 6330 Cham, Switzerland

If disposing of this product, please recycle the paper.

Preface

Institutions are organisational structures that govern the behaviour of individuals within a given community. They essentially specify the 'rules of the game' that coordinate interactions. Electronic institutions are the computational analogue that coordinates the interactions of software agents. Electronic institutions have gained ground in the design and implementation of open multiagent systems, as they allow the specification of the rules of the interaction, and agents can then interact in an autonomous way while complying with the specified rules.

This book introduces electronic institutions, where an institution is represented as a network of scenes, connected by transitions. Just as there are meetings in human institutions in which different people interact, electronic institutions have similar structures, the scenes, to facilitate interactions between agents. These scenes are essentially group meetings, and each scene is defined by an interaction protocol that regulates the agents' interactions in that scene. Protocols are defined for agent roles instead of individual agents. This allows different agents to play the same role in an institution (such as the role of a bidder in an auction), and it also allows the same agent to play different roles in the same interaction (for example, one may be a seller of one item and a buyer of another). The transitions between scenes then describe how agents are allowed to move from one scene to another.

One of the main aims of this book is to illustrate a number of real-life applications of electronic institutions, especially focusing on mixed human/software agent communities. In particular, we present three applications: the uHelp, the WeCurate, and the PeerLearn platforms. In uHelp, a number of community members (mostly parents) volunteer to help each other by offering their services when needed, such as picking up someone's child from school, or baby sitting. An electronic institution is designed to regulate the behaviour of the individuals in this community. It specifies who can request help (for example, it may state that one cannot request any help if one has not volunteered to help others in the last six months), who may be asked to volunteer (for example, only members of the community that are trusted by the person issuing the request are asked for help), how someone's action updates their reputation (for example, if a volunteer is late in picking up a child, then their reputation in picking up children should decrease), and so on.

In WeCurate, a number of museum visitors collectively browse an art gallery online in order to collectively choose a set of 'interesting' objects. An electronic institution is then designed to specify when and how the visitors are allowed to express their interest in an item, how the group's interest as a whole is calculated, how the visitor can argue about or discuss an item in order to influence the group's interest, and when and how the visitors can vote on an item.

In PeerLearn, an online-learning platform, a number of students join a tutor's online classroom. An electronic institution is designed to allow the tutor to provide exercises for the students to perform, and allows the students to practice the subject and upload their work for it to be discussed, commented on, and rated by other students. The electronic institution specifies the rules of the interaction, and it integrates a 'collaborative assessment model' and 'automated evaluation web services' (specific to the music domain) that aid the tutor in assessing the students.

Electronic institutions were initially designed to regulate fixed agent interactions. However, as electronic institutions have become more and more popular for regulating open mixed human-agent communities, the PeerFlow system has been designed and built for peer-to-peer electronic institutions. In the PeerFlow system, an intuitive user interface allows nontechnical users to specify institutions and share them with the community. To keep the specification nontechnical, a mechanism is implemented for automatically building the graphical user interface needed for a human user to interact with a running institution, based on the specification of the institution. For example, if the specification states that the human user should upload a file at a given point in the interaction, then when this point is reached, the automatically generated user interface will provide the user with the appropriate interface for uploading their file. With people sharing their electronic institution specifications, any user may then login, select an existing institution and run it, or alternatively, join an existing running institution. The last part of this book presents the PeerFlow architecture and its automated graphical user interface.

Barcelona,   *Nardine Osman*
December 2023

Acknowledgements

This book is the result of the substantial effort put in by various members of the Artificial Intelligence Research Institute (IIIA-CSIC) at Barcelona over the years, and work on electronic institutions has been supported throughout these years mostly by the Spanish government and the Artificial Intelligence Research Institute. The very early work on electronic institutions started in 1996, motivated by implementing the Blanes fish market as a multiagent interaction. However, work on electronic institutions has grown ever since to become one of the pillars of multiagent interactions, and it has in fact been the seed for the International Workshop on Coordination, Organizations, Institutions and Norms in Agent Systems (COIN), which has run successfully since 2006.

This is the first book to introduce electronic institutions, and with this gentle introduction, we present a number of interesting real-life applications, along with a new direction of electronic institutions: peer-to-peer electronic institutions. However, this work would not have been possible without the effort of many IIIA members, especially Carles Sierra and his three (then) Ph.D. students Pablo Noriega, Juan Antonio Rodríguez-Aguilar, and the late Marc Esteva, along with his (then) Master's student Bruno Rosell. Carles, currently the director at IIIA, remains the main force driving innovative implementations and applications of electronic institutions. Pablo (currently an Ad Honorem researcher at IIIA) and Juan Antonio (currently a Professor at IIIA) have provided extensive and crucial contributions to electronic institutions over the years. Bruno is still the engineer at IIIA who makes electronic institution applications happen. However, our sincere gratitude and respect goes to the late Marc Esteva, whose vast and dedicated research on electronic institutions has been influential in growing and promoting this line of work.

The work on Electronic Institutions has also benefited from the funding of European projects COMRIS (ESPRIT LTR 25500), SLIE (IST-1999-10948), ACE (ACE ERA-Net), and PRAISE (388770), as well as Spanish national projects SMASH (TIC96-1038-C04001), eINSTITUTOR (TIC2000-1414), Agreement Technologies (CSD 2007-0022), CBIT (TIN2010-16306), and uHelp-D (EUIN2015-62530).

Contents

Part II Applications of Electronic Institutions

Part I
Introduction to Electronic Institutions

Chapter 1
A Naive View of Electronic Institutions

Pablo Noriega

> *"Institutions are the rules of the game in a society or, more formally, are the humanly devised constraints that shape human interaction".*
>
> D.C. North [28, p. 3]

This chapter explains the intuitions behind electronic institutions. The first objective is to introduce the ideas and terminology that are the foundations of their conceptual model. The second objective is to give an indication of the usefulness of these systems. For these purposes a particular traditional institution — the fish market that still takes place in fishing villages in Europe — is used to show how we elaborated the notion of electronic institution and also to illustrate how the electronic version of that conventional institution may be specified. Finally, this chapter contains some generic indications of how electronic institutions may be used and some examples that complement the ones that constitute the rest of this book.

1.1 The Fish Market Metaphor

Twice a day in many fishing villages, in Spain and around the world, the village fleet's catch is auctioned in a fish market following a time-honoured tradition. Figure 1.1 depicts an actual auction in the port of Blanes on the Catalonian coast. In spite of its apparently folkloric features and deceptive simplicity, the fish market is not just a *place* (the "llotja" building with boxes, clock, bidding devices and screen) where goods are exchanged under a peculiar *downward-bidding* auction protocol,

---

Pablo Noriega

IIIA-CSIC, Barcelona, e-mail: `pablo@iiia.csic.es`

© Springer Nature Switzerland AG 2024

N. Osman (ed.), *Electronic Institutions*, Artificial Intelligence: Foundations, Theory, and Algorithms, https://doi.org/10.1007/978-3-319-65605-2_1

nor just an *organisation*: the fishermen's guild that runs the auctions. A closer look reveals also an *institution* where goods are traded under exquisitely refined socially acknowledged conventions.

The fish market — like other traditional institutions — serves the important social purpose of enabling several individuals to accomplish an activity that they could not accomplish individually. The fish market and other institutions enable that collective activity by establishing and enforcing artificial ("humanly devised") constraints that articulate the interactions of those agents [28, 29]. In spite of its obviousness, it is worth acknowledging that institutions are also a *virtual social space* where certain facts "count as" something else than what they are outside of that environment. Searle puts it like this: "*An institution is any collectively accepted system of rules (procedures, practices) that enable us to create institutional facts.*" [36, p 21].[1]

This book is about a particular type of institution: *electronic institutions* that were conceived to play the same conventional institutional role of establishing and enforcing artificial constraints to articulate agent interactions. The not-so-subtle difference is that electronic institutions are meant to articulate interactions that may happen online and may involve humans as well as artificial entities.

We take advantage of the actual fish market to motivate and illustrate what are electronic institutions, because auctions are an excellent everyday example of classical institutions and because, in fact, the fish market inspired the conception and development of electronic institutions.

Figure 1.1 shows the afternoon session of the Blanes fish market in the late 1990s.[2] Except for the fact that in this picture some characters (the auctioneer and some buyers) hold electronic communication devices and fish is displayed in plastic boxes, the scene is the same you would have seen in the very same port one hundred years earlier. This scene will serve to illustrate most of the institutional aspects that inspired our conceptual model for electronic institutions. In the upper left corner of the picture, one can see three buyers who seem to be inspecting a particular box at their feet while other buyers (all holding little black boxes) as well as sellers and bystanders complete the cast. All are placed around the sixty or so boxes that make up the day's catch of a single boat. Notice that behind them there are many more boxes; these belong to different ships of the fleet and will be auctioned later on.

The picture captures the moment when the auctioneer (the man in the centre, holding a microphone and a little black box) is about to start a new bidding round. He is looking away from the floor, towards a display board where the information about the next batch of boxes to be auctioned appears. Once he pushes a button of his black box the bidding clock where the unit price is displayed will start ticking down frenetically until one buyer stops it by clicking his own device. What is key in this setup is that everyone present *shares the same state* of the bidding: the boxes, the information on the display board, the running price and the opportunity to click is exactly the same for everyone. Moreover, what is institutionally more significant is

---

[1] See also [35, 18] for a more thorough and formal discussion of how that social space is created. In a similar way electronic institutions provide a computational environment for a virtual social space (see below Sec. 1.3).

[2] This picture belongs to the personal collection of the auctioneer, Albert Ros.

Fig. 1.1 Bidding in the Blanes fish market

that the actions that may take place and the outcomes of any action that takes place in that shared state are known to everyone; because while an auction is taking place, the *rules that govern* every action are *explicit*, always *apply* and are always *enforced* by the auctioneer and the fish market organisation as a whole.

This is the essence of a classical institution that we want to capture in electronic institutions: a collective activity where agents perform within a shared state of affairs, and under the effective enforcement of the explicit rules of the game.

How exactly we capture that essence is described formally and in full detail in [13], but the ideas that spawned that framework took inspiration from several "institutional features" that we identified in the Blanes fish market and we believe are common to most conventional institutions.

Notion 1 *The core* institutional features *that underlie the idea of an electronic institution are:*

1. *Trading is like a play (a collective activity), where actors perform their roles in successive* scenes *according to a script.*
2. *That* script *defines what actors do in each scene (a Dutch auction protocol is used for price determination) and how they move from one scene to another (first sellers get their fish admitted; then, fish may be auctioned; and eventually buyers pay and sellers get paid).*
3. *The script involves a set of entities (like props and stage decor) that are needed for performing the play. Likewise, the fish market contains a set of (standardised) objects, categories and concepts that are involved in the auctioning process:*

boxes to put single types of fish in; a bidding clock, its speed and step; the list of types of fish that are sold; the way fish boxes are labelled or the information that is displayed in each bidding round.

4. *All interactions may be construed as messages.* [3]

5. *All actors in a given scene share the* same *state of the scene (as illustrated with Figure 1.1).*

6. *All agents are subject to constraints that affect those messages and the movement of people. These constraints may be of three types:*

- *Most constraints are* regimented, *(i) as physical constraints; for example boxes of certain size, with labels (for each box) that display standardised information (the standard name of the contents, its quality and its weight, as well as the name of the boat that caught it and so on); (ii) because they are accomplished only through communication devices (bids, bidder identity, clock ticks), or (iii) enforced directly by the fish market staff; for instance, the* auctioneer *declares a bid invalid when a bidder hits the button by accident, or the way that the* receptionist *accepts and labels incoming boxes.*
- *Some constraints are* discretionary *(left to staff to decide when they are enforced); for example, accept an unsupported bid from a reliable customer, breaking bid ties.*
- *Finally, other constraints are difficult for the market to enforce (for example, no deals should be closed outside the market sessions) and compliance with them relies on social norms and commercial regulations that are outside the scope of the institution.*

7. *These conventions are meant to guarantee that the market is* effective *and* trust-worthy *by being* reliable *(stable operation that fosters predictable demand and supply conditions, clear rules, transparency and accountability, proper operation of technological and physical facilities),* safe *(correct accounting, assurance that information and identities are not tampered with, etc.) and* fair *(neutral with respect to buyers and sellers, adequate risk allocation, prudent business model, etc.).*

Since we were all the time interested in making electronic institutions work online, we introduced some additional assumptions about the participants in an *electronic* institution.

Notion 2 *Participants in an electronic institution are* agents *that:* [4]

1. *may be human or artificial;*

---

[3] Recall that during bidding rounds, every action that the buyers and auctioneer perform in order to trade fish is performed through their electronic devices. Moreover, all other actions that take place in the fish market that involve buyers, sellers and staff (for instance, open a credit line, request payment, get charged for a purchase) could be performed through on-site screens or online.

[4] Note that we are implicitly assuming the existence of two different types of entities: namely the "institutional environment" — which constrains actions and keeps a shared social state — and "agents" — those entities that act within that environment and are subject to the constraints imposed by it.

2. *may be active in more than one scene at a time (for example, placing remote bids in several auctions using a phone);*
3. *act on their own will and thus: (i) enter and leave the institution or a scene when they wish, (ii) decide when to act (or not) when they wish, (iii) may be incompetent or malicious;*
4. *the institution does not know the reasons behind the actions of agents;*
5. *agents cannot be forced by the institution to act in any way, but the institution may regiment those actions that may take place; and*
6. *agents can make commitments within the institution and may be held responsible for them outside the institution.*

We took the above features and the points we made about agents, and turned them into the design assumptions of the electronic institution *conceptual model*. Thus the first two institutional features we translated into what we call the *performative structure* consist of *scenes* and *transitions*. Since we think of scenes as an exchange of messages in a restricted language (Notion 1, features 3 and 4) we decided to put all the elements needed to express those messages into what we called the *dialogical framework* and Notion 1, feature 6 gave rise to what we called the *rules of behaviour*. Thus, based on these three main components (dialogical framework, performative structure and rules of behaviour), we may specify — using a *specification language* — an electronic institution. That electronic institution is implemented — on an *electronic institutions platform* — and runs, as illustrated in Figure 1.2, within a larger technological environment that, alongside other systems and agents, works in a given organisational and social environment.

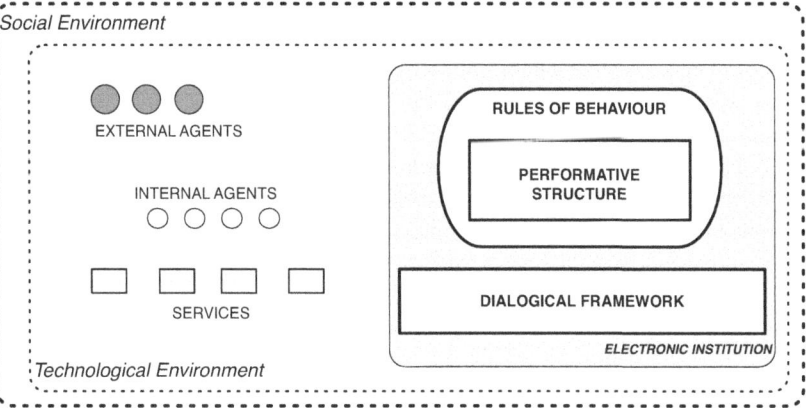

Fig. 1.2 An electronic institution and its context

In this chapter we will use the specification language ISLANDER [15] to show how one can define an electronic institution that implements an online auction house mirroring the fish market.

1.2 An Electronic Institution for the Fish Market

1.2.1 Fencing the Open Field: The Dialogical Framework

The key function of an institution is to create an "institutional world" that is different from the "rest of the world". The point is that, as we saw in the fish market example, only actions that comply with the institutional constraints may have an effect in the institutional world and, likewise, some actions that happen in the institutional world do have an intended effect in the real world. For that purpose one needs to establish a legitimate and explicit correspondence between entities and actions in the real world and entities and actions within the institution. Only when this correspondence is properly established can one claim that an action in the world has institutional meaning, and therefore changes the institutional state and, conversely, that a change in the institutional state has an effect in the world.

Technically, one needs to achieve two goals. First, establish the "constitutive norms" that create a *legitimate institutional reality* and establish the validity of some *institutional facts and actions* as opposed to *brute* facts and actions that exist in the *real world* (see [36]). Second, *make explicit which entities* of the real world are involved in institutional realities, so that institutional actions can take place and reflect their institutional and real-world effects. While the process to achieve the first goal is strictly "pragmatic", in the sense that it happens outside of the specification of the electronic institution and usually involves contracts, legislation, organisations, firms and so on, the second one impinges directly on the specification of an electronic institution.

There are two consequences of these separate realities.

The immediate consequence is that, in order to specify an electronic institution, one needs to make explicit which entities of the real world are involved in institutional realities. We call the set of names that correspond to those institutional entities the "domain ontology" of the institution and we will include all of them in electronic institutions as terms in what we call a *domain language* of the institution. There are also some institutional entities that do not correspond directly to brute facts or actions but are needed for organising collective activities within the institution. We group them in two classes: the *social model* and the *information model* of the electronic institution.

The social model reflects the roles that may be played within an electronic institution and the relationships these roles keep. We distinguish between "internal" roles, those played by agents that are part of the institution — auctioneer, accountant and receptionist in the fish market — and "external" roles, which are played by agents that are not — buyers, sellers and bystanders in the fish market. We allow roles that specialise other roles. We also distinguish between a *static separation* of roles (no individual may play more than one of the separate roles during an enactment of the institution) and *dynamic separation* (an individual may eventually play multiple roles during an enactment but not simultaneously); in the fish market specification the role customer has a static separation into buyer and seller roles. In the fish market we have three external roles: one is `customer` and it subsumes two incompatible roles `buyer` and `seller`. The internal roles are `auctioneer`, `receptionist`, `accountant` and `gatekeeper`, all of them dynamically compatible and all of them subsumed by the role `staff`.

The *information model* is the set of data structures and variables that are involved in the specification and operation of the electronic institution. These include: (i) *institutional terms* that are specific to the application domain and the specification and remain constant during enactments. For example, the list of fish types, the length of timeouts, and the function that calculates the price decrement during a bidding round, (ii) *institutional variables* that capture the evolution of an institution during its enactments, These are either (a) structural (every institution has them), like the number of active scenes, their names, the number of agents in the institution; or (b) performance indicators (specific to the objectives and peculiarities of a given institution, for example the evolution of prices over a period of time). (iii) *scene variables* that capture the activity that takes place in a scene, such as the number and identities of agents in the scene, the agents that are entering or leaving, the contents of a specific bid, the number of invalid bids, the average number of clock ticks before a valid bid. (iv) *transition variables* capture activity within a transition such as the lag between transits, the proportion of roles that visit it, etc. (v) *Agent variables*, such as the number of invalid bids it has submitted, its remaining credit, the scenes where it is active, the time spent in each scene, and so on. (vi) The institutional state that captures all the information that holds at each step of an enactment and includes the current value of all the previous data structures.

The other consequence of the separate institutional reality is that we need to make clear how agents in the real world are going to interact so that they may affect the institutional reality. In other words, we need to adopt an institutional interaction model. Notice that in the fish market, bidding rounds can all be carried out by using electronic communication devices and a display board. As a matter of fact, all institutional interactions within the fish market may be reified as messages between agents. Thus in electronic institutions we choose to make all interactions among agents *utterances* that will be expressed as a sort of speech act in a *communication language*, and organise such utterances as a sort of *dialogue* that we call *scenes*. And that is why we lump the syntactic elements that are needed to express dialogues — languages, social and information models — into one formal construct of the electronic institution: the *dialogical framework*.[5]

---

[5] There is a set of nested languages: domain, communication, action, constraints. At the end of this section there are examples of all of them.

1.2.2 Organising Interactions: Performative Structure

1.2.2.1 Scenes

As we have just discussed, the auctioning of fish is achieved in what may be described in "theatrical" terms as a "bidding rounds" *scene*. There is a cast of characters, a script performed by individuals who perform the established roles, as well as the props and stage — consisting of a clock, a display board, electronic devices and auction hall — that support the performance. We mirror these elements in what we define as a *scene* of an electronic institution.

The "script" of a scene in electronics institutions is defined as a "protocol". In that protocol we express how agents may move from a starting state to successive valid institutional states in order to achieve their individual goals; for instance, to buy fish. As we suggested above, only institutionally valid actions may change an institutional state, and all institutional interactions are expressed as utterances. Consequently, the scene protocol is naturally represented as a finite state graph, whose nodes are institutional states that are linked by utterances. So for example, Figure 1.3 shows the actual specification of the bidding rounds scene we have been talking about so far.

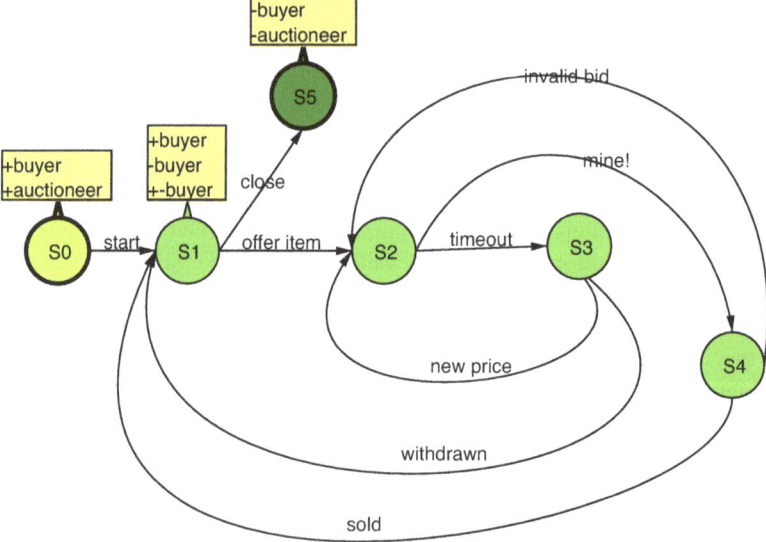

Fig. 1.3 The fish market bidding process (ISLANDER specification of the Auction scene)

In the Blanes fish market , the scene starts when the auctioneer takes his place in the auction room and calls for attention, at which point the display board and communication devices become operative and buyers and public may take their places; next a bidding round starts when the auctioneer points to a box and the corresponding data is displayed on the screen. The auctioneer then states the starting price (clicking his device) and the clock starts displaying lower prices at a fixed rate until either a buyer stops it (by crying "mine" or clicking her device), or a "reservation price" is reached.

The same process is described in Figure 1.3. The bidding rounds become active when the *auctioneer* declares start — moving from state S0 to state S1 — and then points to an item and offers it at a starting price (S2). Next, if no buyer places a bid before the clock takes a step timeout, the goods are offered at a next lower new price or withdrawn if the new price is below or equal to the reservation price; but if a *buyer* cries mine! before the clock takes one step, the auctioneer either declares the box sold or declares the bid invalid because there is an anomaly — a tie or an "accidental" bid. When an anomaly is detected, the same box is put up for auction with a new starting price (usually 20% above the anomalous bid). When a reservation price is reached, or a box is sold, the box is withdrawn from the auction (we are back in S1) and the next auctionable box is put up for bids until the last one is sold or withdrawn.

A scene protocol, like the script of a play, also specifies when actors enter and leave the scene. Notice that Figure 1.3 has some boxes attached to some states. These are states where agents that play a given role may enter (noted by "+") or leave ("−") the scene. A well defined scene has to have at least one entry state for each role involved and at least one exit state that can be reached from entry states. Notice also that there may be states where a role is marked with a "+ -" pair; this is a way of enabling agents to be active, simultaneously, in more than one scene. For instance in S1 the *buyer* role is labelled with a plus (any buyer can enter then), a minus sign (any buyer inside the auction room may leave then) *and* a plus and minus pair, which means that an agent playing that role may move to another scene but still remain in the current scene. In this case a buyer may continue bidding in that scene but may join other auction rooms or update her credit.[6]

1.2.2.2 Performative Structure

Is there more than one scene in the fish market?, you may ask. Yes. There is a scene Reception that deals with the "reception of goods". That is, before fish can be sold, the catch of each boat is brought into the llotja and put into boxes with specimens of the same species and quality. These boxes are then inspected, weighed and labelled by a llotja employee (the *receptionist*) before all the boxes of that ship are brought into the auction room and displayed for potential buyers to inspect. There is another scene — Admission — where buyers validate their identities and set up credit lines with the auction house and yet another one — Accounting — where buyers may

---

[6] Scenes, as we said, can be understood as conversations. It makes sense even in conventional institutions that one person may participate in two conversations, or be active in two auction halls (for instance, bidding in several sea ports through a phone and a representative [12]).

update their credit, pay for purchases and get their purchases delivered, and sellers get their accounts settled.

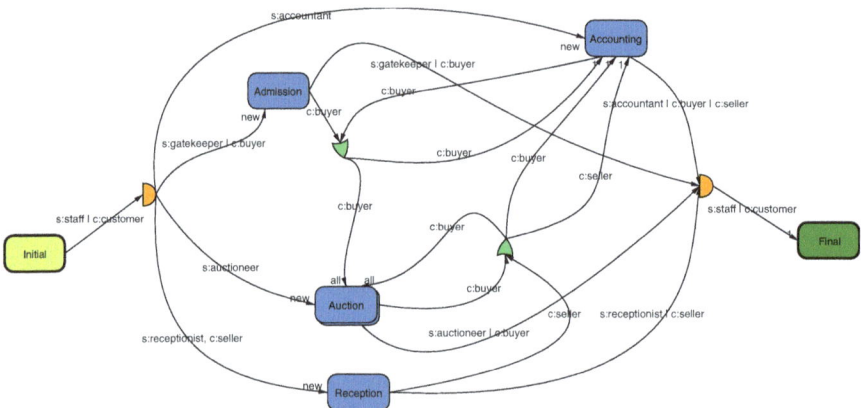

Fig. 1.4 The FM performative structure

Figure 1.4 is the actual specification of the fish market performative structure we have been describing. It consist of six scenes (boxes) and four transitions (arrow-like icons). There are two scenes (Initial and Final) that are common to every institution and serve to activate and deactivate agents that enter and leave the electronic institution. The other four (substantive) scenes: Admission, Accounting, Auction and Reception, serve to coordinate interactions of a group of agents towards the common goal of completing sales. Note that in this specification, the Auction scene is *multiply instantiated*; this means that (i) several bidding rounds may take place (simultaneously or not), (ii) agents may be active in zero, one or several of these scenes (simultaneously or not), and (iii) each auction scene has its own bidding protocol (which may be different for any scene — Dutch, English, Vickrey, whatever).

1.2.2.3 Transitions

The arrow-like icons in Fig 1.4 are *transitions* whose function is to prescribe how agents *performing given roles* may move from one *scene* to another (in the direction indicated by the arrow-like icon). Transitions are connected to a scene by a single *incoming* arc and may connect to $N \geq 1$ (not necessarily distinnct) scenes through *N outgoing* arcs. Notice that scene Admission has two connections leading towards Auction and Final. Auction has three connections coming from Initial and Admission and also from itself (since an agent may change auctions and may be active in several auctions), while scene Reception has one in and two out (Accounting and Final).

There are the two types of transitions: AND transitions — semicircle sign — where each role must follow one path; transitions are of the same type, and OR — empty arrow-like boxes — meaning an agent may take any possible destination.

Arcs are labelled with *agent-id:role* pairs indicating that agents playing those roles may move from one scene to another, for instance agents playing the roles of `auctioneer` or `buyer` may move from Auction to Final although only a `buyer` may also move from Auction to Accounting. Agents may change roles "inside a transition" (as long as they are compatible) — for example s enters as `staff` and leaves as `auctioneer` between Initial and Auction.

Notice that outgoing arcs have another label that indicates whether that role is creating a "new" instance of a scene — the *auctioneer* opens Auction — or may enter either "1" (one) or "all" (any) such instances. A transition may also contain constraints that state the minimum or maximum number of agents that are needed for the scene to become or remain open and to withhold entry within capacity, for instance Auction may not be opened unless there are at least thirty buyers in the transition from the Admission scene.

A last comment: performative structures may have other performative structures embedded. In this case transitions connect to the initial and final scenes of the (sub) performative structure.

1.2.3 Rules of Behaviour

One can understand the performative structure as a way of expressing procedural conventions. A scene is specified as an interaction protocol and transitions state the conventions to proceed from one scene to another. However one should be aware that because of the way scenes and transitions are specified, they also involve *functional and behavioural conventions* and in fact they *regiment* them. We discuss the "normative" content of scenes, but it should be clear that the same ideas apply to transitions.

We said that a scene is specified as a finite state machine and the link between two states of the scene is labelled by an utterance that involves a speaker that plays a role. However, the specification of each link includes, one one hand, the preconditions that need to hold in the electronic institution at the moment of the utterance so that the potential speaker can actually "speak" and, on the other, the post-conditions that specify the effects that a successful utterance will have in the institution. The specification thus gives the agent "permission" to speak as long as it satisfies the preconditions but the specification effectively forbids the agent to speak if it does not satisfy those preconditions, thus regimenting a functional rule. The post-conditions alter the state of the institution, thus regimenting also a functional rule of behaviour.

For example in the Auction scene protocol, the actual utterance that joins states S3 and S2 (labelled "new price" in Figure 1.3) is in fact the following illocutionary schema:

```
(inform (|x auctioneer) (all buyer) offer ?price)
```

where !x, the last agent that played the role of auctioneer, performs the action of informing all those agents present that are playing the buyer role that the price (of the item being auctioned) that will apply will be the value of the variable ?price. However this schema has one precondition (constraint): the new price should be larger than the reservation price:

$$!reservation > !price - decrement$$

Unless this condition holds, the auctioneer will not be permitted to utter a new offer. On the other hand, if the precondition holds, the effect of the utterance (or the action that takes place) will be that the value that is assigned to the variable price is the current price minus the value of the decrement:

$$?price == !price - decrement$$

1.2.4 Constitutive Norms

We suggested above (Sect. 1.2.1) that if one expects to make the electronic institution version of the fish market legitimate, one needs to address pragmatic issues. The electronic institution is a convenient way to specify how interactions should take place and to ensure that they take place according to this specification. However, the actual implementation of a conventional institution involves the existence of an organisation that applies those conventions and is responsible for its proper application. The fish market has been successful over the centuries by defining and enforcing stable conditions on four main aspects: (i) acceptable behaviour of participants within the site; (ii) the availability, presentation and delivery of goods; (iii) the eligibility requirements for participating buyers and sellers; and (iv) the satisfaction of public commitments made by participants. The first one is achieved with the electronic institution, while the other three involve constitutive conventions that are satisfied in the Blanes market *organisation* and would need minor tailoring to apply in an online operation. A similar situation applies to other electronic institutions.

1.3 Four Ways of Understanding Electronic Institutions

Mimetic perspective: **electronic institutions as computational systems that perform the same function performed by conventional institutions.**

We mentioned in Sect.1.1 that classical institutions — such as fish markets, a college fraternity, a professional sports league — can be seen as sets of artificial constraints to articulate agent interactions around a common goal. The previous section shows how one may go about building such electronic counterparts. The aspect that is relevant is that such counterparts (i) allow one to articulate new forms of interactions: those that involve human and artificial agents in a hyper-connected society, and (ii) enforce the conventions that govern interactions in an automated fashion.

Governance perspective: **electronic institutions as open regulated multiagent systems.**

Electronic institutions create a special type of restricted environment where interactions among agents take place. The environment plus the participating agents can be seen as a multiagent system (MAS). Because of the assumptions on agents we stated in Notion 2, the MAS can be qualified as an open MAS., and because of institutional features (Notion 1) 6 and 7 and (Notion 2) agency assumption 5, the MAS may be properly referred to as "regulated". One may even claim that electronic institutions may be properly called "normative MASs" [2], especially when compared with the discussion in [23]. The contentious issue, if any, is whether one accepts that non-regimented enforcement and online scene specification are compatible with the *EI* model, and appropriate middleware and tools are available.[7]

Artefact perspective: **electronic institutions as operational interfaces between the subjective decision-making processes of individuals and a collective task achieved through their interactions.**

H. Simon (in [37]) refers to markets as an interface between the individual goals of buyers and sellers that enables them to exchange goods. The market is not concerned with the way individuals reach a decision, it merely exists — as an independent third party — to facilitate the goal of buying and selling goods or services. Likewise, in general, the electronic institution mediates between the goals and motivation of individuals within an activity that they may not achieve individually.

Note that this view of electronic institutions is akin to the ideas of mechanism design and, in particular, quite natural for exploratory or simulation-based mechanism design, but it applies as well to other forms of collective problem solving or social coordination such as collective decision making and opinion aggregation.

Coordination support: **the conceptual model of electronic institutions as a metamodel for social coordination.**

The conceptual model of electronic institutions may be seen as a "metamodel" that affords some generic means to achieve coordination in a rather general type of hybrid online social systems. It is a metamodel in the sense that the same *EI* constructs may be used to specify (or model) a wide variety of particular institutions, each of which

---

[7] These matters are discussed in some detail in [13] and in [24].

serves to support coordination of particular collective activities. As such, electronic institutions brings its peculiar set of abstractions, analogies and tools to the challenge of affording social intelligence.

1.4 The Use of Electronic Institutions

1.4.1 A Brief History of Electronic Institutions

The origin of electronic institutions was the realisation — in the early days of agent technologies and the Internet (1995) — that one should pay distinct attention to the context where agents are situated. If one wanted to use the rationality encapsulated in autonomous agents to do anything other than push or pull information from web pages, one needed to address the problem of meaningful interaction. We addressed that problem by focusing on the type of activity we thought would be the simplest possible one both in terms of a minimal domain ontology involved and a minimal interaction.

We chose auctions, which turned out to be a fortunate choice; not especially because, as we naively presumed, one could build on a sound analytic understanding of them ([20, 40]), but because they are not trivial, they are successful examples of actual institutions, they were promising as a mechanism for electronic commerce and, however superficially, everyone is acquainted with them. Moreover, we chose the Blanes fish market in particular because (i) it exhibits all the features that a real auction house has, (ii) we had immediate access to it and we could therefore analyse its operation in full detail, (iii) it ought to be a good example, because it has been in operation since at least the fourteenth century, (iv) it is picturesque and would provide "marketing" advantages, and (v) its downward-bidding protocol had coordination complexities that were technically challenging. And this turned out to be a fruitful decision because what started as an interest in one particular type of auction evolved into the realisation that other forms of collective coordination shared the institutional features that we identified in the Blanes fish market.

This inspired the first conceptualisation of a "dialogical institution" with the three main components of the current "electronic institutions" conceptual model, which together with an application of electronic institutions for argumentation-based negotiation, are the core of [22], supervised by Carles Sierra and Ramón López de Mántaras. J.A. Rodríguez-Aguilar, in his dissertation [33], systematised the existing notions, developed the notion of transition to its current conception and produced tools to specify and run electronic institutions, while M. Esteva's dissertation [14] refined the definitions and produced the architecture and the full suite of tools that constitute the *EIDE* platform [16], including the ISLANDER specification language [15]. Both dissertations were also under the supervision of Carles Sierra. Almost a decade after the original ideas were put in print, with a few more theses on the subject written and several applications developed, we finally came up with a complete formal description of the electronic institutions conceptual model (*EI*) in [13]. The ideas are still going strong as the examples of this book show and new dissertations such as [19] prove.The group that has been involved in the development of these ideas is large and it is not easy to acknowledge all contributions but the bibliography includes over forty names of those collaborators and an indication of their involvement.

The *EI* model was extended in [17] to include a normative language to express scene protocols and allow for non-regimented norms and their governance. A different expansion came with the addition of a layer for 3-D representation of electronic institutions [38] and, on top of this, with the addition of software assistants and other means to make interfaces more ergonomic [39, 1]. Another strand of extensions has been to endow electronic institutions with means of evolution [5, 9, 10]. More recently, there has been a line of development that strives to have simpler, more agile electronic institutions. This is the case of *PEERFLOW* with a specification language, *SIMPLE* — which, unlike ISLANDER, does not include ubiquity of agents, nor run-time creation of scenes, nor complex transitions between scenes, but has more intuitive graphic conventions — and a peer-to-peer architecture that allows distributed execution of electronic institutions.

1.4.2 The Usefulness of Electronic Institutions

From the perspective of applications design, one can identify some distinctive features of electronic institutions that make them valuable both as a conceptual model and as a framework for deploying socio-technical systems. The main four are:

1. Electronic institutions are concerned with the design and implementation of the environment and not with the internals of participating agents. This fosters:

    a. a sharper awareness of coordination and governance features;
    b. a separation of concerns between the design of social conventions and the design of the capabilities of individuals; and
    c. an encapsulation of best practices and procedural conventions through scenes and scene networks that capture them.

2. Electronic institutions implement open systems in which external agents:

    a. may enter and leave the institution at any labelled entry or exit state of a scene;
    b. may be independent (built or owned independently of the institution);
    c. are self-motivated (have their own goals); and
    d. have their own internal model not controlled by the institution.

3. The electronic institutions conceptual model is neutral with respect to the agent architecture, thus facilitating the participation of both human and computational agents — in debugging or enactment — if adequate interfaces are provided.

4. Scenes and scene networks foster modular design and re-usability. In particular, scenes and role changes between scenes may be individually edited. Consequently, scene variants are simple to implement and available scenes may be pruned or spliced into a network.

5. The electronic institution model allows the coexistence of several active scenes at any time and, more specifically, the possibility for any agent to be active, simultaneously, in more than one. Each scene is subject to its particular conventions but the electronic institution model guarantees that commitments are duly propagated among scenes.

1.4.3 An Overview of Applications

Over the years, the electronic institutions framework has been used to support a wide variety of applications.

Generally speaking, electronic institutions have been successfully used in four major types of application:

- Regulated systems where their intended or natural use or their debugging involve humans and agents in a symbiotic fashion. For instance participatory simulation, online dispute resolution and online multiplayer games.
- Interactive collective coordination systems that need robust and reliable governance by a trusted third party because of liability and risk associated with interactions. Typical example are electronic trading, voting and election markets, gambling or grid computing.
- Regulated systems where a full reliable modelling, debugging or monitoring with conventional analytic or software engineering tools is not sufficient or even feasible, for instance because all potentially undesirable consequences cannot be found analytically and some form of complex experimentation is needed, as in some forms of mechanism design, or because exploration of consequences and adaptation of conventions is difficult to achieve at design-time with conventional tools, such as the case of pubic policy design and management.
- Coordination support systems where procedural rules need to be flexible or adaptable to unforeseen situations and artificial agents may profitably be used as support to users and some users may themselves be software agents. Such is for instance the case with activities that involve workflows that are constantly changing or where several variants of some protocols may need to become available at some point during the system's lifespan.

In addition to the examples included in the rest of this book, a concise but systematic description of several applications of the *EI/EIDE* framework, prior to 2012, is included in [13]. Those applications include, among others, electronic markets for fish and for electricity [12, 22, 4], agent-based simulation for archaeology and policy-making [7, 11, 8], online games [3], automated negotiation [31], online dispute resolution [27] and grid computing [6], procurement, supply networks and open innovation [21] and flexible work-flow specification for hotel management systems [32] and for scientific assessment [30].

1.5 Further Reading

1. This paper includes a complete formal specification of the electronic institutions model, together with a revision of related work:

- Mark d'Inverno, Michael Luck, Pablo Noriega, Juan A. Rodríguez-Aguilar, and Carles Sierra. Communicating open systems. *Artificial Intelligence*, 186(0):38 – 94, 2012.

2. For additional information about related work, this book compiles the description of eight frameworks for social coordination, one of which is electronic institutions. Its contents are organised with the idea of facilitating a comparison of the strengths and weaknesses of the frameworks:

- Huib Aldewereld, Olivier Boissier, Virginia Dignum, Pablo Noriega, and Julian Padget. *Social Coordination Frameworks for Social Technical Systems*. Number 30 in Law, Governance and Technology Series. Springer International Publishing, 2016.

The last part of Chapter 4 includes a critical assessment of the design assumptions of electronic institutions:

- Pablo Noriega and Dave de Jonge. Electronic institutions: The EI/EIDE framework. In Huib Aldewereld, Olivier Boissier, Virginia Dignum, Pablo Noriega, and Julian Padget, editors, *Social Coordination Frameworks for Social Technical Systems*, pages 47–76. Springer International Publishing, Cham, 2016.

2. Finally, these four papers [25, 34, 22, 26] show the early understanding we had of auctions and electronic institutions.

References

1. Almajano, P., Mayas, E., Rodriguez, I., Lopez-Sanchez, M., Puig, A.: Structuring interactions in a hybrid virtual environment - infrastructure & usability. In: Proceedings of the International Conference on Computer Graphics Theory and Applications and International Conference on Information Visualization Theory and Applications (VISIGRAPP 2013), pp. 288–297 (2013). DOI 10.5220/0004215802880297
2. Andrighetto, G., Governatori, G., Noriega, P., van der Torre, L.W.N. (eds.): Normative Multi-Agent Systems, *Dagstuhl Follow-Ups*, vol. 4. Schloss Dagstuhl - Leibniz-Zentrum fuer Informatik (2013)

3. Aranda, G., Trescak, T., Esteva, M., Carrascosa, C.: Building quests for online games with virtual institutions. In: F. Dignum (ed.) Workshop on Agents for Games and Simulations at AAMAS 2010, *Lecture Notes in Computer Science*, vol. 6525, pp. 192–206. Springer, Berlin (2010)

4. Arcos, J.L., Noriega, P., Rodríguez-Aguilar, J.A., Sierra, C.: E4MAS through electronic institutions. In: D. Weyns, H. Parunak, F. Michel (eds.) Environments for Multi-Agent Systems III, no. 4389 in Lecture Notes in Computer Science, pp. 184–202. Springer, Berlin (2007)

5. Arcos, J.L., Rodríguez-Aguilar, J.A., Rosell, B.: Engineering autonomic electronic institutions. In: D. Weyns, S. Brueckner, Y. Demazeau (eds.) Engineering Environment-Mediated Multi-Agent Systems, *Lecture Notes in Computer Science*, vol. 5049, pp. 76–87. Springer, Berlin (2008)

6. Ashri, R., Payne, T., Luck, M., Surridge, M., Sierra, C., Rodríguez-Aguilar, J.A., Noriega, P.: Using electronic institutions to secure grid environments. In: M. Klusch, M. Rovatsos, T. Payne (eds.) Cooperative Information Agents X, *Lecture Notes in Computer Science*, vol. 4149, pp. 461–475. Springer, Berlin (2006)

7. Bogdanovych, A., Simoff, S.: Establishing social order in 3d virtual worlds with virtual institutions. In: A. Rea (ed.) Security in Virtual Worlds, 3D Webs, and Immersive Environments: Models for Development, Interaction, and Management, pp. 140–169. IGI Global, Hershey (2011)

8. Botti, V., Garrido, A., Giret, A., Noriega, P.: The role of MAS as a decision support tool in a water-rights market. In: F. Dechesne, H. Hattori, A. ter Mors, J. Such, D. Weyns, F. Dignum (eds.) Advanced Agent Technology, *Lecture Notes in Computer Science*, vol. 7068, pp. 35–49. Springer, Berlin (2012). URL http://dx.doi.org/10.1007/978-3-642-27216-5_4

9. Bou, E., Lopez-Sanchez, M., Rodríguez-Aguilar, J.A.: Adaptation of autonomic electronic institutions through norms and institutional agents. In: G. O'Hare, A. Ricci, M. O'Grady, O. Dikenelli (eds.) Engineering Societies in the Agents World VII, *Lecture Notes in Computer Science*, vol. 4457, pp. 300–319. Springer, Berlin (2007)

10. Campos, J., Lopez-Sanchez, M., Esteva, M.: Using a two-level multi-agent system architecture. In: M. De Vos, N. Fornara, J.V. Pitt, G. Vouros (eds.) Coordination, Organization, Institutions and Norms in Agent Systems VI (COIN 2010), *Lecture Notes in Computer Science*, vol. 6541. Springer, Berlin (2011)

11. de la Cruz, D., Estévez, J., Noriega, P., Pérez, M., Piqué, R., Sabater-Mir, J., Vila, A., Villatoro, D.: Norms in H-F-G societies. Grounds for agent-based social simulation. In: F. Contreras, M. Farjas, F. Melero (eds.) CAA'2010 Fusion of Cultures. Proceedings of the 38th Conference on Computer Applications and Quantitative Methods in Archaeology. Archaeopress, Oxford (9998). There is a Spanish version in Cuadernos de Prehistoria y Arqueología, No. 20; pp 149-161., 2010 (U. Granada)

12. Cuní, G., Esteva, M., Garcia, P., Puertas, E., Sierra, C., Solchaga, T.: MASFIT: Multi-agent systems for fish trading. In: 16th European Conference on Artificial Intelligence (ECAI 2004), pp. 710–714. IOS Press, Amsterdam (2004)

13. d'Inverno, M., Luck, M., Noriega, P., Rodríguez-Aguilar, J.A., Sierra, C.: Communicating open systems. Artificial Intelligence **186**(0), 38–94 (2012). DOI 10.1016/j.artint.2012.03.004. URL http://www.sciencedirect.com/science/article/pii/S0004370212000252

14. Esteva, M.: Electronic Institutions: from specification to development. Ph.D. thesis Universitat Politècnica de Catalunya (UPC), 2003. No. 19 in IIIA Monograph Series. IIIA (2003)

15. Esteva, M., de la Cruz, D., Sierra, C.: ISLANDER: an electronic institutions editor. In: Proceedings of the First International Joint Conference on Autonomous Agents and Multiagent systems (AAMAS '02), pp. 1045–1052. ACM Press, New York (2002)

16. Esteva, M., Rodríguez-Aguilar, J.A., Arcos, J.L., Sierra, C., Noriega, P., Rosell, B.: Electronic institutions development environment. In: Proceedings of the 7th International Joint Conference on Autonomous Agents and Multiagent Systems (AAMAS '08), pp. 1657–1658. International Foundation for Autonomous Agents and Multiagent Systems, ACM Press, New York (2008)

17. Garcia-Camino, A.: Normative Regulation of Open Multi-Agent Systems. Ph.D. thesis Universitat Autònoma de Barcelona, 2010. No. 35 in IIIA Monograph Series. IIIA (2011)
18. Jones, A., Sergot, M.: A formal characterization of institutionalized power. Logic Journal of the IGPL **4**(3), 427–446 (1996)
19. de Jonge, D.: Negotiations over Large Agreement Spaces. Ph.D. thesis Universitat Autònoma de Barcelona, 2015. IIIA Monograph Series. IIIA (In press)
20. McAfee, R.P., McMillan, J.: Auctions and bidding. Journal of Economic Literature **XXV**, 699–738 (1987)
21. Montero, R., de la Cruz, D., Noriega, P.: Prototipo de una plataforma de negociación on-line para el mercado de residuos. tr-iiia-2013-02. Tech. rep., IIIA - CSIC, Barcelona (2013)
22. Noriega, P.: Agent-Mediated Auctions: The Fishmarket Metaphor. Ph.D. thesis Universitat Autònoma de Barcelona, 1997. No. 8 in IIIA Monograph Series. IIIA (1999)
23. Noriega, P., Chopra, A.K., Fornara, N., Cardoso, H.L., Singh, M.P.: Regulated MAS: Social Perspective, chap. 4, pp. 93–133. No. 4 in Dagstuhl Follow-Ups. Schloss Dagstuhl–Leibniz-Zentrum fuer Informatik (2013). URL http://drops.dagstuhl.de/opus/volltexte/2013/4001
24. Noriega, P., de Jonge, D.: Electronic institutions: The EI/EIDE framework. In: H. Aldewereld, O. Boissier, V. Dignum, P. Noriega, J. Padget (eds.) Social Coordination Frameworks for Social Technical Systems, pp. 47–76. Springer International Publishing, Cham (2016). DOI 10.1007/978-3-319-33570-4_4. URL http://dx.doi.org/10.1007/978-3-319-33570-4_4
25. Noriega, P., Sierra, C.: Towards layered dialogical agents. In: J.P. Müller, M.J. Wooldridge, N.R. Jennings (eds.) Third International Workshop on Agent Theories, Architectures, and Languages, ATAL-96, *Lecture Notes in Computer Science*, vol. 1193, pp. 173–188. Springer, Berlin (1996)
26. Noriega, P., Sierra, C.: Auctions and multi-agent systems. In: M. Klusch (ed.) Intelligent Information Agents: Agent-Based Information Discovery and Management on the Internet, pp. 153–175. Springer, Berlin (1999). DOI 10.1007/978-3-642-60018-0_9. URL http://dx.doi.org/10.1007/978-3-642-60018-0_9
27. Noriega, P., López de Toro, C.: Towards a platform for on-line mediation. In: M. Poblet, U. Shild, J. Zeleznikow (eds.) Proc. Workshop on Legal and Negotiation Decision Support Systems (LDSS 2009) in conjunction with ICAIL 2009, pp. 67–75. CEUR Workshop Proceedings, Barcelona (2009)
28. North, D.C.: Institutions, Institutional Change and Economic Performance. Cambridge University Press, Cambridge (1990)
29. North, D.C.: Institutions. Economic Perspectives **5**(1), 97–112 (1991)
30. Osman, N., Sierra, C., Sabater-Mir, J., Wakeling, J.R., Simon, J., Origgi, G., Casati, R.: LiquidPublications and its technical and legal challenges. In: D. Bourcier, P. Casanovas, M. Dulong de Rosnay, C. Maracke (eds.) Intelligent Multimedia: Managing Creative Works in a Digital World, vol. 8, pp. 321–336. European Press Academic Publishing, Florence (2010)
31. Ramchurn, S.D., Sierra, C., Godo, L., Jennings, N.R.: Negotiating using rewards. Artificial Intelligence **171**(10-15), 805–837 (2007). DOI 10.1016/j.artint.2007.04.014. URL http://www.sciencedirect.com/science/article/pii/S0004370207000756
32. Robles, A., Noriega, P., Cantú, F.: An agent oriented hotel information system. In: K.S. Decker, J.S. Sichman, C. Sierra, C. Castelfranchi (eds.) Proceedings of the 8th International Confonference on Autonomous Agents and Multiagent Systems (AAMAS '09), pp. 1415–1416. International Foundation for Autonomous Agents and Multiagent Systems, ACM Press, New York (2009)
33. Rodríguez-Aguilar, J.A.: On the Design and Construction of Agent-Mediated Electronic Institutions, Ph.D. thesis, Universitat Autònoma de Barcelona, 2001. No. 14 in IIIA Monograph Series. IIIA (2003)
34. Rodríguez-Aguilar, J.A., Noriega, P., Sierra, C., Padget, J.: FM96.5 a Java-based electronic auction house. In: Proceedings of the Second International Conference on The Practical Application of Intelligent Agents and Multi-Agent Technology (PAAM'97), pp. 207–224. The Practical Application Company, London (1997)

35. Searle, J.R.: The Construction of Social Reality. Free Press, New York (1995)
36. Searle, J.R.: What is an institution? Journal of Institutional Economics **1**(01), 1–22 (2005). DOI 10.1017/S1744137405000020. URL http://dx.doi.org/10.1017/S1744137405000020
37. Simon, H.A.: The Sciences of the Artificial, third edn. MIT Press (1996)
38. Trescak, T., Esteva, M., Rodríguez, I.: VIXEE an innovative communication infrastructure for virtual institutions. In: Proceedings of the 10th International Joint Conference on Autonomous Agents and Multiagent Systems (AAMAS 2011). ACM Press, New York (2011)
39. Trescak, T., Rodríguez, I., López-Sánchez, M., Almajano, P.: Execution infrastructure for normative virtual environments. Engineering Applications of Artificial Intelligence **26**(1), 51–62 (2013). URL http://www.sciencedirect.com/science/article/pii/S0952197612002540
40. Wolfstetter, E.: Auctions: an introduction. Journal of Economic Surveys **10**(4), 367–420 (1996)

Chapter 2
An Abstract View of Electronic Institutions

Carles Sierra and Pablo Noriega

The purpose of this chapter is to put in precise terms the intuitions presented in the previous one. So, let us start with an abstract version of what conventional institutions are.

Notion 3 *An* institution *is an artefact that creates a virtual social* (institutional) *space, inside of which* agents *interact under clearly established "rules of the game". This entails the following features:*

f.1 a boundary *that distinguishes when agents are inside or outside the institutional space and when they interact within that space or outside of it;*

f.2 an explicit connection *between the institutional and the external space;*

f.3 a restricted institutional ontology, *so that inside the space, actions can only involve a class of objects, actions, properties, facts, events, etc. that is common to all participants and is the one involved in all the activity that happens inside the institutional space;*

f.4 a common interaction model, *so that agents* can *interact amongst themselves and with the space itself, and in fact* may *interact only through this institutional interaction model;*

f.5 an explicit collection of conventions *(constraints, rules, norms, etc.) that apply to all participants and are properly enforced inside the institutional space;*

f.6 a persistent (but changing) state of the social space that (i) is the same for all participants and (ii) may change through the actions of participants in accordance with the institutional conventions.

An *electronic institution* is a computational version of this characterisation that interprets these six features in a particular way. In fact we will characterise electronic

Carles Sierra
IIIA-CSIC, Barcelona, e-mail: `sierra@iiia.csic.es`

Pablo Noriega
IIIA-CSIC, Barcelona, e-mail: `pablo@iiia.csic.es`

© Springer Nature Switzerland AG 2024
N. Osman (ed.), *Electronic Institutions*, Artificial Intelligence: Foundations, Theory, and Algorithms, https://doi.org/10.1007/978-3-319-65605-2_2

institutions as a class of socio-cognitive technical systems — systems that exhibit some features akin to those in Notion 3 and are essential for electronic institutions.

So, this chapter starts with a generic characterisation of socio-cognitive technical systems and makes explicit in rather intuitive terms what is distinctive about the class of electronic institutions (Sect. 2.1). The next sections give a precise characterisation of the electronic institutions conceptual framework (the *EI metamodel*). Sec. 2.8 presents one electronic institutions platform, the *Electronic Institutions Development Environment* (from now on, the *EIDE* platform), which implements the *EI* metamodel, and Sec. 2.9 contains a brief discussion of some of its variants.

2.1 Socio-cognitive Technical Systems and Electronic Institutions

In the last chapter we mentioned that the fish market as a *conventional institution* involves three complementary entities:

1. $\mathcal{W}$ the *fragment of the real world* that is involved in the trading of fish. Namely, the organisation owned by the fishermen's guild that sells fish, pays taxes, hires auctioneers and signs contracts with buyers and sellers.

2. $\mathcal{I}$ the *institution* itself consisting of the conventions and rules of how fish is traded that are enforced by that organisation.

3. $\mathcal{T}$ the *technology* — consisting of the bidder and auctioneer remote control devices used by buyers and auctioneer, the clock and display boards and so on — that makes it work.

In the fish market *electronic institution*, the relationship between $\mathcal{W}$ and $\mathcal{T}$ plays a more significant role (i.e. the interaction model in *f.4* Notion 3 is necessarily in $\mathcal{W}$). The point is that in electronic institutions any action in the world that may have an institutional effect has to be performed through an input to the computational system (using the system interfaces) and all effects of institutional actions become apparent only through the interfaces in $\mathcal{W}$.

This tripartite view (the "WIT-trinity") applies to the class of "socio-cognitive technical systems" (SCTSs), of which electronic institutions is a subclass. Hence, let us first define what we mean by SCTS and then look into the tripartite relationship in some detail.[1]

Notion 4 *A* Socio-cognitive technical system (SCTS) *[27] is a multiagent system that satisfies the following assumptions:*

A.1 *System A socio-cognitive technical system is composed of two ("first-class") entities: a* social space *and the* agents *who act within that space. The system exists in the real world and there is a boundary that determines what is inside the system and what is outside.*

---

[1] We use the term SCTS as a simplifying label. They have been called "socio-technical" [34], *socio-cognitive systems* [11], *intelligent socio-technical systems* [25] and *hybrid online social systems* in [27].

A.2 *Agents Agents are entities who are capable of acting within the social space. They exhibit the following characteristics:*

    A.2.1 *Socio-cognitive Agents are presumed to base their actions on some internal decision model. The decision-making behaviour of agents, in principle, takes into account social aspects because the actions of agents may be affected by the social space or other agents and may affect other agents and the space itself [11].*

    A.2.2 *Opaque The system, in principle, has no access to the decision-making models or internal states of participating agents.*

    A.2.3 *Hybrid Agents may be human or software entities (we shall call them all "agents" or "participants" where it is not necessary to distinguish).*

    A.2.4 *Heterogeneous Agents may have different decision models, different motivations and respond to different principals.*

    A.2.5 *Autonomous Agents are not necessarily competent or benevolent, hence they may fail to act as expected or demanded of them.*

A.3 *Persistence The social space may change either as an effect of the actions of the participants, or as an effect of events that are caused (or admitted) by the system.*

A.4 *Perceivable All interactions within the shared social space are mediated by technological artefacts — that is, as far as the system is concerned there are no direct interactions between agents outside the system and only those actions that are mediated by a technological artefact that is part of the system may have effects in the system — and although they might be described in terms of the five senses, they can collectively be considered percepts.*

A.5 *Openness Agents may enter and leave the social space and it is not known a priori (by the system or other agents) which agents may be active at a given time, nor whether new agents will join at some point or not.*

A.6 *Constrained In order to coordinate actions, the space includes (and governs) regulations, obligations, norms or conventions that agents are in principle supposed to follow.*

This characterisation of SCTSs captures a large class of actual systems but has the advantage of making explicit some elements that fit the tripartite understanding and are essential in electronic institutions. The tripartite view (the WIT trinity) is interesting because it elucidates the key features in Notion 3 and serves to make precise how, in particular, they are addressed in electronic institutions. Figure 2.1 sketches the main ideas.[2]

---

[2] We choose the term "trinity" to stress the fact there is one single SCTS that may be viewed in three different ways; each of these views has its own characteristic features but the three views are interrelated in an indissoluble way and constitute *one* SCTS. Notice that this is a static view of an SCTS; a sequence of trinities is needed to represent the evolution of an SCTS. Moreover this trinity does not reflect the fact that SCTSs are embedded in an ecosystem where other SCTSs as well as several legislative or regulatory frameworks and technological artefacts may be involved [12].

Notice first that the relationships between the three views are denoted by three double-headed arrows. (i) Thus the arrow that joins $\mathcal{W}$ and $\mathcal{I}$ addresses features *f.1–3* of Notion 3. It establishes the legitimacy and meaningfulness of the "institutional reality" we discussed in the last chapter. One direction is the "counts-as" function [33, 24], and the other is the "anchoring" of the terms of the domain language in crude entities.

(ii) The downward arrow between $\mathcal{I}$ and $\mathcal{T}$ states that the implementation is correct (the implementation behaves the way it is intended to behave) and the upwards arrow holds if the institutional constraints are a model of the implementation; that is, when the institutional conventions happen to prescribe the way the system actually works (feature f.5).

(iii) Finally, the arrow that joins $\mathcal{W}$ and $\mathcal{T}$ addresses features *f.1* and *f.4*. It means that the interface works properly because (a) artefacts (input devices, network, etc.) work as designed, (b) information is not corrupted going into or out of $\mathcal{T}$ and (c) all the information that is relevant for the operation of the system flows into the system and only information that is relevant flows in and out.

Feature *f.6* is achieved when we claim that the three arrows establish a sort of "isomorphism" between the *shared state* of each of the three views, as illustrated in Figure 2.2.

The intuition is that at any point in time, an (institutionally relevant) action in the world changes the world, if and only if its institutional version changes the institutional state according to the institutional conventions and the same happens in the computational state. In other words, that the *state of the world*, as far as the system is concerned, changes if and when an *attempted* action in $\mathcal{W}$ is *validated* by $\mathcal{I}$, and then the code in $\mathcal{T}$ *processes* the input that happened in $\mathcal{W}$ and — because $\mathcal{T}$ is correctly programmed according to $\mathcal{I}$ — $\mathcal{T}$ make the effects available as outputs in $\mathcal{W}$.

Although this is ongoing work [28], to make the idea of such an "isomporphism" more precise one needs to address two objectives. Firstly, from the perspective of an agent, to *afford* it the means to be aware of the state of the world in order to

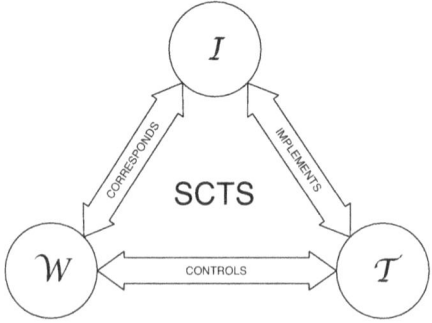

Fig. 2.1 The tripartite view of socio-cognitive technical systems: The ideal system ($\mathcal{I}$), the technological artefacts that implement it ($\mathcal{T}$) and the actual organisation that uses the system ($\mathcal{W}$) (after [27])

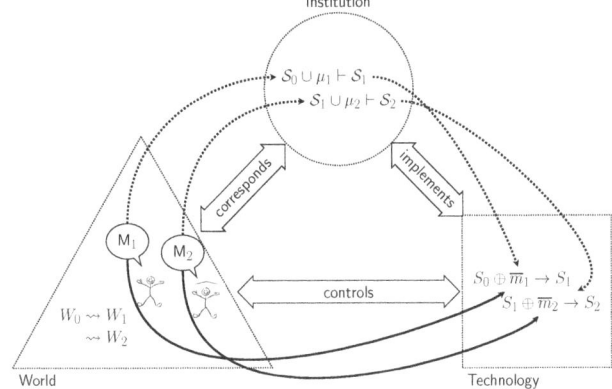

Fig. 2.2 Shared context in an SCTS ([28])

decide whether to act or not; and, in order to act, to have the elements to coordinate its actions with other agents within the social space. Secondly, the social space, to fulfil those needs — as suggested in Notion 3 — has to establish ontology, boundary, interaction model and effective governance. In the end, these are the functionalities that we need to be able to model, specify and implement in an SCTS. In other words, we need a *metamodel* that provides the formal constructs needed to achieve those "affordances" of a social space, and a technological *platform* for turning that specification into actual artefacts.

We put this in simpler terms as follows.

Notion 5 *An* affordance *(of the social space of an SCTS) is a property of the social space that supports effective interactions of agents within an SCTS.*

In the previous paragraph we have implicitly postulated three *affordances* that are essential in every SCTS:

1. *awareness*, which provides participating entities access to those elements of the shared state of the world to enable them to decide what to do,
2. *coordination*, so that the actions of individuals are conducive to the collective endeavour that brings them to participate in the SCTS, and
3. *validity*, which preserves the proper correspondences of the tripartite view.

We distinguish these because they contribute directly first, to the establishment of individual perception of (common) social situations, second to the realisation of the mechanisms for collective action and third to the correctness of the activity as a whole. However, other affordances are needed in order to support the features in Notion 2.

Note that *awareness*, *coordination* and other affordances as well may be achieved by a variety of means. Hence, one needs to make explicit the particular "constructs" (languages, data structures, operators, operational semantics etc.) through which these properties are achieved in a given SCTS. Since the same constructs may be used to model different SCTSs, we will use the notion of "metamodel" to capture a collection of such constructs; and because some metamodels may fit a particular SCTS better than others, we will link a given SCTS with a particular instantiation of a given metamodel.

Notion 6 A metamodel *(for SCTSs) is a collection of languages, data structures and operations that when instantiated produce a model of an SCTS (and its internal agents, if any), through features that achieve the affordances of awareness, coordination — and perhaps others — in a social space.*

Finally, a model is simply a "good" description of a socio-cognitive system that captures all its intended affordances.

Notion 7 A model *of an artificial socio-cognitive system S is an instantiation of a metamodel for SCTS, such that the view of S in W matches the view of S in I.*

Note that this "matching" entails that the integrity requirements of the three relationships in the WIT trinity are in fact correctly achieved. In particular (i) the *counts-as* relationship is correctly established by participants having the proper entitlements and an appropriate bijection between terms in I and objects and potential actions in W, (ii) the model is faithfully implemented in T and (iii) the input/output flow between T and W is not corrupted.

A platform — a set of technological artefacts — is meant to provide such *faithful implementations* and therefore make the WIT trinity loop *valid.*

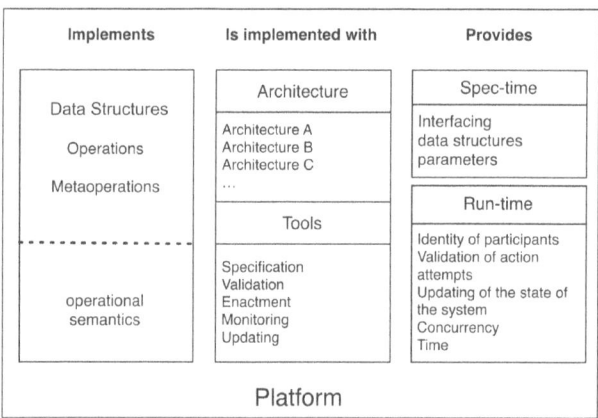

Fig. 2.3 Contents and functions of an SCTS platform [28]

Notion 8 A platform *is a set of technological artefacts that implement the affordances of a metamodel in order to specify, implement and support the execution of an actual SCTS model S, so that the view of S in T corresponds to the view of S in I.*

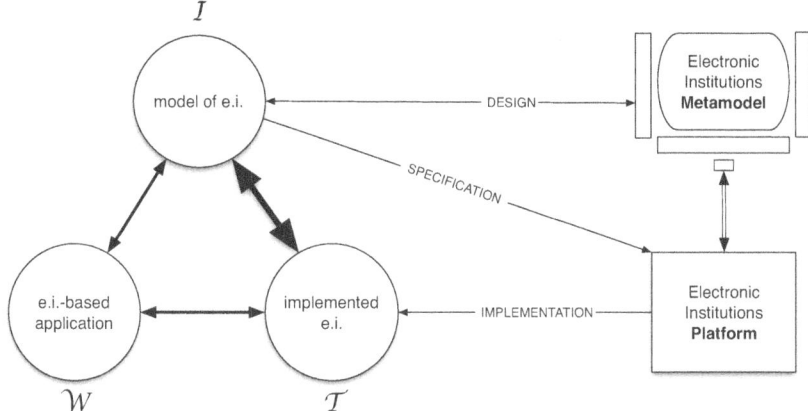

Fig. 2.4 The WIT-trinity understanding of electronic institutions

As Figure 2.3 illustrates, a platform contains two sorts of objects, on one hand a computational architecture that will host the implemented SCTS, on the other a set of software tools that are used to specify, debug, update, run and monitor the actual SCTS. When an SCTS is enacted on a platform, the platform provides an infrastructure or environment that supports the SCTS at run-time.

2.1.1 Electronic Institutions as a Subclass of Socio-cognitive Technical Systems

We may now postulate what we mean by electronic institutions (Figure 2.4).

Notion 9 *An electronic institutions is a member of the class of socio-cognitive technical systems that are modellable with an electronic institution metamodel and implemented with an electronic institution platform.*

In the next sections we shall give a rather precise definition of a particular electronic institutions metamodel, EI, which corresponds to the "conceptual framework" we used in the previous chapter to describe the fish market. Before doing so, we will comment on its main assumptions and the affordances that distinguish EI and its variants from other SCTSs.[3]

Notion 10 *The* EI metamodel affords:

1. An institutional space *that is*

---

[3] That EI metamodel and the corresponding EIDE platform have been used to implement most of the applications mentioned in the previous chapter, although some EI-compatible variants were used for example in [29, 31, 2, 7] and in the next chapters.

- Open: *Agents are self-motivated, heterogeneous black boxes and may enter and leave the institutional space at their own will. Neither the infrastructure nor the participants know* a priori *what external agents may become active in an enactment of an electronic institution or scene.*
- Persistent: *There is a* restricted ontology *of only those objects, relationships, functions and actions that are expressible in the* domain language *used for labelling utterances. All institutional actions are acknowledged by the institution during an enactment.*
- Social: *The institution supports collective endeavours and agents are ascribed* roles *within a social model.*

2. A mode of interaction: *Institutional effects are achieved through* dialogical *interactions construed as point-to-point messages in a communication language.*
3. Coordination patterns: *The totality of interactions are organised as* activities *that are:*

- Decomposable *into interdependent simpler activities* (scenes) *that achieve particular goals with the participation of fewer individuals.*
- Contextual: *The scope of interactions is limited. Those subgroups of agents that act within a scene share a common "scene context" that gives rise to a* state of the scene *that is common to all agents within that scene. The institution as a whole constitutes a "global context" that holds for every agent (where the states of all scenes are joined into a unique coherent* institutional state*).*
- Replicable: *Scenes may be either re-enacted by different groups of agents or enacted concurrently with different groups.*
- Co-incident: *An agent may be active, simultaneously, in more than a single activity.*

4. Governance: *Constraints are dealt with in two fashions. Some are "regimented" as part of the institutional specification — through the three main constructs discussed below — and therefore enforced by the electronic institution infrastructure, while others are embedded in the decision-making models of internal agents and are enforced — possibly with some laxity — through a "law-enforcement" action of those internal agents. Constraints are* role-based.

These affordances are supported with the three core constructs that we mentioned in the last chapter: dialogical framework, performative structure and norms. We will discuss them rigorously in the next section but some intuitive remarks may be appropriate.

Notion 11 *The EI metamodel is based on the following* core elements:

- *Agents and Roles.* Agents are the players in an electronic institution, interacting by the exchange of illocutions, whereas roles are defined as standardised patterns of behaviour. The identification and regulation of roles is considered to be part of the formalisation process of any organisation [32]. Any agent within an electronic institution is required to adopt some role(s). A major advantage of using roles is that they can be updated without having to update the actions for every agent on an individual basis. Recently, the concept of role is increasingly being studied by software engineering researchers [30, 23], and by researchers in the agents community [5, 26, 35, 9, 6]. Hereafter we will differentiate between *internal* (institutional) and *external* (non-institutional) roles as well as internal, sometimes called staff, and external agents. Internal roles are those enacted to carry out the institution's services and tasks, and to guarantee some of the institutional rules. Hence, only staff agents are allowed to adopt internal roles, while external agents are those playing external roles.

- *Dialogical framework.* Some aspects of an institution such as the objects of the world and the language employed for communicating are fixed, constituting the context or domain of interaction amongst agents. In a dialogical institution, agents interact through speech acts (illocutions). Institutions establish the acceptable speech acts by defining the ontology and the common language for communication and knowledge representation, which are bundled in what we call the dialogical framework. By sharing a dialogical framework, we enable heterogeneous agents to exchange knowledge with other agents.[4]

- *Scene.* Interactions between agents are articulated through agent group meetings, which we call *scenes*, with a well-defined communication protocol. We consider the protocol of a scene to be the specification of the possible dialogues agents may have. Notice however that the communication protocol defining the possible interactions within a scene is role-based instead of agent-based. In other words, a scene defines a role-based framework of interaction for agents.

- *Performative structure.* Scenes can be connected, composing a network of scenes that we call the performative structure, to capture the relationships among scenes. The specification of a performative structure contains a description of how agents can legally move from scene to scene by defining the pre-conditions both to join and leave scenes. Satisfying such conditions will fundamentally depend on the roles allowed to be played by each agent and its acquired commitments. The execution of a performative structure equates to the execution of multiple, possibly simultaneous, ongoing activities, represented by scenes, along with the agents participating in each activity. Agents within a performative structure may be participating in different scenes, with different roles, and at the same time.

- *Normative rules.* Agent actions, in the context of an institution, have consequences, shaped as compromises which impose obligations or restrictions on dialogical actions of agents in the scenes wherein they are acting or will be acting in the future. The purpose of these normative rules is to affect the behaviour of agents by imposing obligations or prohibitions. Obligations are understood as illocutions that must be uttered and prohibitions as illocutions that must not be uttered. This basic view of norms permits the coding of more sophisticated representations of norms while being computationally simple. Enforcement of what must not be said can be easily put in place by filtering out illegal illocutions

---

[4] Here we adhere to the definition of Alberts in [1]: *An ontology for a body of knowledge concerning a particular task or domain describes a taxonomy of concepts for that task or domain that define the semantic interpretation of the knowledge.*

and the detection of violations of what had to be said but was not said is also straightforward.

The EI metamodel is defined in the next sections (and summarised in Figure 2.5). A particular EI platform, EIDE, is briefly discussed in Sect. 2.8

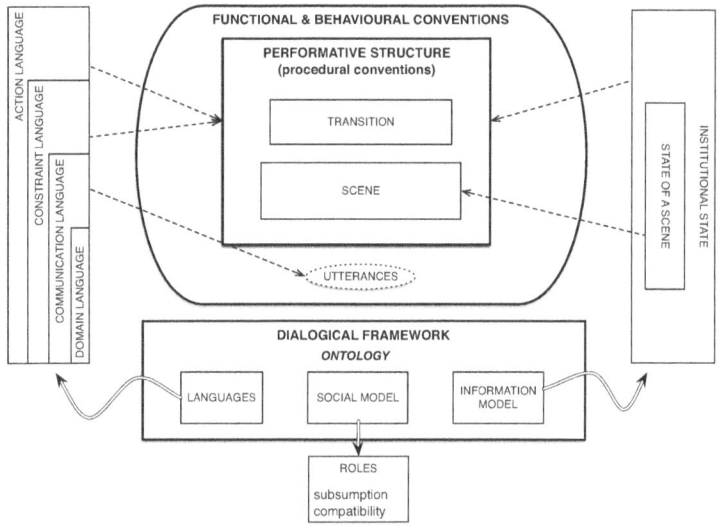

Fig. 2.5 An outline of the EI metamodel (after [18])

2.2 Dialogical Framework

The Dialogical Framework establishes the necessary conventions to allow heterogeneous agents to communicate, and to express the procedural conventions that govern their interaction. Notice that in the most general case, each agent immersed in a multiagent environment is endowed with its own inner language and ontology. Hence, in order to allow agents to successfully interact with each other we must address the fundamental issue of putting their languages and ontologies in relation. For this purpose, the dialogical framework fixes communication, constraint and action languages and establishes the institutional ontology and its social structure.

Definition 2.1. We define a *dialogical framework* as a tuple $df = (o, \Gamma, \Lambda, sm, im)$ where

- o stands for an ontology;
- Γ is a set of illocutionary particles;

- Λ is a set of languages;
- *sm* is a social model;
- *im* is an information model.

2.2.1 Ontology

First of all and for notational purposes we introduce some definitions that will apply henceforth. Let V be a set of variables, K a set of constants and $\Upsilon = V \cup K$ a set of symbols. By V_i, K_i and Υ_i we denote the sets of variables, constants and symbols of type i. These sets contain constants and variables of the domain as conceptualised in the ontology, along with constants and variables of the electronic institution. For instance, V_{agents} denotes the set of agent variables and K_{scene} the set of scene names (identifiers). In the formalisation uppercase letters denote sets, while lowercase letters denote elements.

An ontology is a formal conceptualisation of a domain. Thus, the ontology of an electronic institution contains the formalisation of the relevant concepts of the domain the institution is modelling and therefore fixes what agents can talk about.

Although there are many competing alternatives to represent ontological knowledge, mostly based on Description Logics, for instance OWL[5] or RDF[6], we take a simple object-oriented approach in the ontology definition that is powerful enough to represent a large variety of problems. Hence, the domain is formalised as a set of classes representing the different domain concepts and a hierarchy relationship between them. We use $B = \{integer, real, boolean, string\}$ as the set of basic data types and allow for the definition of enumerated types, which are defined as finite sets of values.

Definition 2.2. We define an ontology as a tuple $o = (E, C, <)$ where:

- $E = \{(e_i, D_i)\}_{i \in I_E}$ is a set of enumerated type definitions where $e_i \in K_{type}$ is the enumerated type identifier and $D_i \subseteq K$ is a set of values.
- $C = \{(c_i, A_i, \rho_{c_i}, \sigma_{c_i})\}_{i \in I_C}$ is a set of class definitions, each one defined as a tuple, where $c_i \in K_{type}$ stands for the class identifier, $A_i \subseteq K_{attrib}$ is a set of attribute identifiers, $\rho_{c_i} : A_i \longrightarrow Bool$ tells which attributes must receive a value when defining a term representing an instance of the class and which attributes may be left unspecified, and $\sigma_{c_i} : A_i \longrightarrow T$ maps each attribute to its type, where T is recursively defined by the following rules:
  - $(B \cup \{e_i\}_{i \in I_E} \cup \{c_i\}_{i \in I_C}) \subset T$
  - if $t_i, t_j \in T$ then $t_i \times t_j \in T$
  - if $t_i \in T$ then $t_i \ list \in T$
  - Nothing else belongs to T.

- $\prec$ is a partial order over K_{class}, which must be regarded as a class hierarchy, such that if $c_i \prec c_j$ then $A_j \subseteq A_i$,

satisfying the requirements below:

1. Enumeration type identifiers should be unique: $\forall i, j \in I_E.(e_i \neq e_j)$
2. Domain values should belong to just one type: $D_i \cap D_j = \emptyset$ if $i \neq j$
3. Class identifiers should be unique: $\forall i, j \in I_C.(c_i \neq c_j)$
4. The class identifiers, enumeration type identifiers and basic data type identifiers must be different: $\{e_i\}_{i \in I_E} \cap \{c_i\}_{i \in I_C} = \emptyset$, $(\{e_i\}_{i \in I_E} \cup \{c_i\}_{i \in I_C}) \cap B = \emptyset$
5. Class inheritance preserves attribute characteristics: if $c_i \prec c_j$ then for all a if $\rho_{c_j}(a)$ is true then $\rho_{c_i}(a)$ is true, and for all a we have $\sigma_{c_j}(a) = \sigma_{c_i}(a)$
6. The class hierarchy does not contain cycles.

While the previous definition establishes how the domain concepts are formalised, agents will exchange illocutions that contain actual instances of the ontology. Next we define how these terms associated with a given ontology are written, that is, how the *instances* of the different concepts are expressed.

Definition 2.3. The set of terms of an ontology $o = (E, C, \prec)$, denoted by $terms$, is recursively defined by the following rules:

- $\{k, k : t | k \in K_t\} \subset terms_t$ for all $t \in B$
- $\{d, d : e | d \in D\} \subset terms_e$ for all $(e, D) \in E$
- $\{c(a_1 = p_1, \ldots, a_n = p_n) | a_i \in A, p_i \in terms_{\sigma_c(a_i)}, \forall a_j \in A.\rho(a_j) \rightarrow a_j \in \{a_1, \ldots, a_n\}\} \subset terms_c$ for all $(c, A, \rho, \sigma_c) \in C$
- $\{v, v : t | v \in V_t\} \subset terms_t$ for all $t \in T$
- $\{(p, q), (p, q) : t_i \times t_j | p \in terms_{t_i}, q \in terms_{t_j}\} \subseteq terms_{t_i \times t_j}$
- $\{[p_1, \ldots, p_n] | p_i \in terms_t, n \geq 0\} \subseteq terms_{t\,list}$ for each $t \in T$
- $nil \in terms_t$ for all $t \in T$
- $terms = \bigcup_{t \in T} terms_t$

We define $terms_t^K \subseteq terms_t$ as the set of terms of type t that do not contain variables. Similarly for $terms^K \subseteq terms$.

2.2.2 Domain Language

We define a domain language that will serve as the basis for the representation languages that will model the behaviour of agents within the context of scenes, namely, communication, constraint and action languages. Before presenting the domain language we define Ω as the set containing the following operations (t stands for any type in the set T):

- Arithmetic operations[7]:

---

[7] These operations are overloaded and are also defined over reals.

– $<, <=, >=, >$: $integer \times integer \rightarrow boolean$
– $+, -, /, \times$: $integer \times integer \rightarrow integer$
– $-$: $integer \rightarrow integer$

- Boolean operations:

  – $\vee, \wedge$: $boolean \times boolean \rightarrow boolean$
  – $\neg$: $boolean \rightarrow boolean$

- Polymorphic List operations:

  – $\in, \notin$: $t \times t\, list \rightarrow boolean$
  – $\cup, -$: $t\, list \times t\, list \rightarrow t\, list$
  – $::, -$: $t \times t\, list \rightarrow t\, list$
  – hd : $t\, list \rightarrow t$
  – tl : $t\, list \rightarrow t\, list$

- Operators to access variable values:

  – $Bindings$: $V_t \times Integer \rightarrow t\, list$
  – $Bindings$: $V_t \times Integer \times K_{state} \times K_{state} \rightarrow t\, list$

- Polymorphic comparisons: $=, \neq$: $t \times t \rightarrow boolean$

Definition 2.4. Given an ontology $o = (E, C, <)$, a set of variables V, a set of constants K and the set of basic operations Ω,[8] we define the domain language $\mathcal{L}_E$ as the language generated by the following grammar with starting symbol E:

$$
\begin{aligned}
E ::=\ & E\, op\, E && \text{with } op \in \Omega_2 \\
& |\, op(E) && \text{with } op \in \Omega \setminus \Omega_2 \\
& |\, p && \text{with } p \in terms \\
& |\, v\,.\,R && \text{with } v \in V \\
& |\, R \\
& |\, E\,,\,E
\end{aligned}
$$

$$
\begin{aligned}
R ::=\ & a && \text{with } a \in K_{Attrib} \\
& |\, R\,.\,a && \text{with } a \in K_{Attrib}
\end{aligned}
$$

By $\mathcal{L}_E^t$ we denote the set of all expressions of type t.

2.2.3 Communication Language

Similarly to other agent communication languages, illocutions in electronic institutions contain an illocutionary particle, expressing the intention when uttering the

---

[8] By $\Omega_2 \subset \Omega$ we denote the set containing the binary operators.

illocution, the sender and addressee(s), the message content, which must be an ontology term, and a time term to capture the particular instant in which an illocution is uttered. The communication language expresses that an illocution is addressed to an agent, to all the agents playing a given role or to all the agents in a conversation, i.e. in a scene. If the illocution is addressed to just one agent this is expressed by a pair containing an agent symbol and a role symbol; if the illocution is addressed to all the agents playing a role this is expressed by a role symbol; and finally, if the message is addressed to all the agents in a conversation this is expressed by the particle "*all*".

Definition 2.5. Given an ontology $o = (E, C, \prec)$, a social model $sm = (R_I, R_E, R_S, \theta, f_I)$, and a set of symbols Γ representing illocutionary particles, we define the communication language $\mathcal{L}_{CL}$ as the language generated by the following grammar with starting symbol CL:

$$CL ::= \iota\,(\,x:r\,,A\,,\varphi:t\,,\tau\,) \quad \text{with } \iota \in \Gamma, x \in \Upsilon_{Agent}, r \in (\Upsilon_{R_I} \cup \Upsilon_{R_E}),$$
$$\varphi \in terms, t \in T, \tau \in \Upsilon_{time}$$

$$A ::= \ (\,x:r\,) \quad \text{with } x \in \Upsilon_{Agent}, r \in (\Upsilon_{R_I} \cup \Upsilon_{R_E})$$
$$\quad |\,r \qquad\quad \text{with } r \in (\Upsilon_{R_I} \cup \Upsilon_{R_E})$$
$$\quad |\,\textbf{all}$$

We say that a communication language expression is an *illocutionary schema* when at least one of the terms contains one or more variable symbols. Otherwise, we say that a communication language expression is an *illocution*. This distinction will be important when specifying scenes.

2.2.4 Constraint Language

In the context of an institution some actions can only be performed when some conditions hold. Hence, in order to capture such conditions, in this section we present a constraint language, which is later on used on the specification of scenes and performative structures. Constraints are specified as a sequence of boolean expressions and list iteration statements. The latter permits us to verify whether a certain condition is satisfied by each element of a list.

Definition 2.6. Given an ontology $o = (E, C, \prec)$, we define the constraint language $\mathcal{L}_C$ as the language generated by the following grammar with starting symbol S:

$$S ::= \ S\ ;\ e \qquad\qquad\qquad\ \text{with } e \in \mathcal{L}_E^{boolean}$$
$$\quad |\,e \qquad\qquad\qquad\qquad \text{with } e \in \mathcal{L}_E^{boolean}$$
$$\quad |\,\textbf{for each } v \textbf{ in } Q \textbf{ do } S \textbf{ endfor} \quad \text{with } v \in V$$

$$Q ::= \quad v \qquad \quad \text{with } v \in V$$
$$\quad \quad | \, v \, . \, R \quad \text{with } v \in V$$
$$\quad \quad | \, R$$

$$R ::= \quad a \qquad \quad \text{with } a \in K_{Attrib}$$
$$\quad \quad | \, R \, . \, a \quad \text{with } a \in K_{Attrib}$$

2.2.5 Action Language

During the institution execution the values of the attributes of the different information models are modified as a consequence of the agents' actions. In order to specify how they are modified, we define a simple language containing assignment, conditional and list iteration statements.

Definition 2.7. Given an ontology $o = (E, C, <)$, we define an action language $\mathcal{L}_A$ as the language generated by the following grammar with starting symbol A:

$$A ::= \quad L = e \qquad \qquad \qquad \text{with } e \in \mathcal{L}_E$$
$$\quad \quad | \, A \, ; \, A$$
$$\quad \quad | \, \textbf{if } e \textbf{ then } A \textbf{ else } A \textbf{ endif} \qquad \text{with } e \in \mathcal{L}_E^{boolean}$$
$$\quad \quad | \, \textbf{if } e \textbf{ then } A \textbf{ endif} \qquad \qquad \text{with } e \in \mathcal{L}_E^{boolean}$$
$$\quad \quad | \, \textbf{for each } v \textbf{ in } Q \textbf{ do } A \textbf{ endfor} \quad \text{with } v \in V$$

$$L ::= \quad v \, . \, R \quad \text{with } v \in V$$
$$\quad \quad | \, R$$

$$Q ::= \quad v \qquad \quad \text{with } v \in V$$
$$\quad \quad | \, v \, . \, R \quad \text{with } v \in V$$
$$\quad \quad | \, R$$

$$R ::= \quad a \qquad \quad \text{with } a \in K_{Attrib}$$
$$\quad \quad | \, R \, . \, a \quad \text{with } a \in K_{Attrib}$$

2.2.6 Information Model

Electronic institutions can model persistent organisations of software that need to keep information about the state of computation. That state may affect the behaviour of participants, that is, the behaviour of agents playing certain roles, for instance in determining the acceptable terms inside illocutions or the satisfaction of constraints in the performative structure. Information must also be kept for monitoring and security reasons. For instance, within an auction house institution the designer may need to maintain information about the credit of each buyer and the list of registered goods for auctioning. We will organise the information to be kept as *information models* associated with the different components of the institution specification. Each information model will be specified as a set of attributes, whose type must be one of the types defined in the institution ontology. The specification of an information model also includes the definition of the default (initial) value of each attribute.

Definition 2.8. Given an ontology $o = (E, C, <)$, we define an information model as a tuple $i = (A, \sigma, \delta)$ where:

- $A \subseteq K_{attrib}$ stands for a set of attribute identifiers;
- $\sigma : A \longrightarrow T$ maps each attribute to its type;
- $\delta : A \longrightarrow terms^K$ returns for each attribute its default value, such that $\delta(a) \in terms^K_{\sigma(a)}$.

2.2.7 Social Model

The notion of role is central in the specification of electronic institutions, allowing us to abstract from the individuals, the agents, that get involved in an institution's activities. This is especially important in open systems whose participants are not known at specification time and can change over time. In this context, each role defines a pattern of behaviour within the institution. Hence, all the actions that can be done within an institution are associated with roles, which can be regarded as agent types. More precisely, we define a role as a finite set of actions. Such actions are intended to represent the set of shared capabilities of the agents that incarnate a particular role. For instance, an agent playing the buyer role must be capable of submitting bids and an agent playing the auctioneer role must be able to offer goods at auction. In order to take part in an electronic institution, an agent is required to adopt some role(s) and to conform to the pattern of behaviour attached to that particular role(s). All agents adopting the same role are guaranteed to have the same rights, duties and opportunities.

From the set of roles that structure an institution, we differentiate between the internal and the external roles. The internal roles are those played by internal or staff agents, which are like the institution employees in conventional institutions. These agents are in charge of guaranteeing the correct execution of an institution. For instance, an auctioneer is in charge of auctioning goods following the specified protocol and a receptionist is in charge of guaranteeing that only agents satisfying the admission conditions for buyers will be allowed to participate with that role within the institution. Usually, the agents incarnating these roles will be programmed by the developers of the institution specification since their behaviour is crucial for the overall correct functioning of the institution. The social model of an institution must allow the specification of relationships among roles, for instance, roles that cannot be played both at the same time. Finally, an information model is associated with each role, establishing what information will be kept by the institution for those agents playing the role.

Definition 2.9. We define a *social model* as a tuple $sm = (R_I, R_E, R_S, \theta, f_I)$ where

- $R_I \subseteq K_{R_I}$ is the set of internal role identifiers;
- $R_E \subseteq K_{R_E}$ is the set of external role identifiers;
- $R_S \subseteq K_{R_S}$ is the set of relationships over role identifiers;
- $\theta : R_S \to \mathcal{P}((R_I \cup R_E) \times (R_I \cup R_E))$ returns for each relationship the set of pairs of role identifiers related by it;
- $f_I : (R_I \cup R_E)-> I$ maps each role to its information model;

satisfying the requirement below:

1. The internal and external roles must be disjoint sets: $R_I \cap R_E = \emptyset$

2.3 Scenes

Recall that the whole activity within an electronic institution was described as a composition of multiple, well-separated and possibly concurrent dialogical activities, each one involving different groups of agents playing different roles. For each activity, interactions between agents are articulated through agent group conversations, which we call *scenes*, which follow well-defined protocols. We consider the protocol of each scene to model the possible dialogical interactions between roles instead of agents. In other words, scene protocols are patterns of multi-role conversation, and any agent participating in a scene has to play one of its roles. It is generic in the sense that it can be repeatedly played by different groups of agents, in the same sense that the same scene in a play can be performed by different actors incarnating the same characters. Scenes represent the context to interpret the uttered illocutions. Context is a fundamental aspect that humans use in order to interpret the information they receive. The same message in a different context may certainly have a different meaning.

A scene protocol is specified by a finite state machine where the nodes represent the different states of the conversation and the labels of the directed arcs contain the actions that make the scene state evolve. The graph has a single initial state (non-reachable once left) and a set of final states representing the different endings of the conversation. There is no arc connecting a final state to any other state.

Because we aim at modelling multiagent conversations whose set of participants may vary dynamically, scenes will allow agents to join in or leave at some particular moments (states) during an ongoing conversation depending on their role. For this purpose, we differentiate, for each role, the sets of access and exit states. Moreover, a set of *stay and go* states are defined for each role. These states permit agents to join new scenes (conversations), without leaving the current one. Normally the correct evolution of a conversation protocol requires that a certain number of agents for each role are involved in it. Thus, a minimum and maximum number of agents per role is defined and the number of agents playing each role has to be always between these limits. This restriction must be taken into account in order to allow agents to either join or leave the conversation. Obviously, the final states have to be an exit state for each role, in order to allow all the agents to leave when the scene is finished. On the other hand, the initial state has to be an access state for those roles whose minimum is greater than zero, in order to start the scene.

The arcs of the finite state machine are labelled with the actions that make the scene evolve at execution time. These are either illocutionary schemes from the communication language in the dialogical framework, or timeouts, which are defined as numeric expressions. While the former define the information agents can exchange, the later permit us to make the scene evolve after a given number of time units have passed since a state was reached. Timeouts are especially important for robustness (to evolve from states where the (external) agents capable of talking and making the conversation progress have been killed or are trying to foot-drag the other agents by remaining silent). An important issue is that if there is an arc from one state to itself, this transition does not stop the timeout countdown. The timeout countdown starts when the state is reached and stops only if there is a transition to a different state of the scene.

In order for protocols to be generic, state transitions cannot be labelled by ground illocutions. Instead, illocutionary schemes must be used where, at least, the terms referring to agents and time must be variables. We should point out that the type of a variable within the context of a scene must be the same for all its occurrences.

Furthermore, each arc may be labelled with a constraint and an action language expression. The constraint language expression must be regarded as the pre-condition of the action that has to be satisfied in order for the action to make the scene state evolve from the source state of the arc to its target state. Constraints can restrict the valid illocutions that agents can utter, as well as the paths that the scene execution can follow. On the other hand, the action language expression defines the consequences of the action. It specifies how information model attributes are modified as a result of the action. Each arc is thus labelled with a set of triplets composed of a constraint, an illocutionary scheme or timeout, and an action language expression. The set has to be interpreted as a disjunction of actions that might make the scene evolve from the source state of the arc to its target state. The first action, i.e. illocution uttered by an agent, that satisfies the constraint in the corresponding triplet will make the scene state evolve.

During a scene conversation, the variables in illocutionary schemes are bound to the matched values of the uttered illocutions. These bindings change dynamically, that is, the same variable may appear in several schemes and thus be bound to different values at different points of the conversation. These bindings may restrict the valid messages in a certain moment (state) of the conversation. That is to say, the same message in the same state may not be valid because the context has changed, i.e. the current variable bindings. For instance, imagine a scene auctioning goods following the English auction protocol. When a valid bid is submitted by a buyer the valid bids for the rest are reduced to bids greater than this one. That is to say, each submitted bid reduces the valid illocutions that buyers can utter, although the scene may remain in the very same state.

In order to verify the correctness of subsequent illocutions the institution infrastructure has to keep track of variable bindings. The expression (domain) language introduced earlier permits us to manipulate the variable bindings in order to express complex constraints over the past moves of the conversation.

For instance, constraints allow us to force the value of a variable in an uttered illocution to be equal to its last bound value by writing: $x = hd(bindings(x, 1))$.[9]

In conclusion, for an illocution uttered by an agent to be valid it has to match an illocutionary scheme of an outgoing arc of the current state, and the constraints associated with this arc have to be satisfied. Then, as the scene reaches the target state of the arc the corresponding action language expression is executed. The next definition summarises the components introduced so far.

Definition 2.10. A *scene type* is a tuple:

$$s = (R, W, w_0, W_f, (WA_r)_{r \in R}, (WE_r)_{r \in R}, (WSG_r)_{r \in R}, \Theta, \lambda, min, Max, i, p)$$

where

- $R \subseteq K_{R_I} \cup K_{R_E}$ is the set of roles of the scene;
- $W \subseteq K_{state}$ is a finite, non-empty set of scene states;
- $w_0 \in W$ is the initial state;
- $W_f \subseteq W$ is the non-empty set of final states;
- $(WA_r)_{r \in R} \subseteq W$ is a family of non-empty sets such that WA_r stands for the set of access states for the role $r \in R$;
- $(WE_r)_{r \in R} \subseteq W$ is a family of non-empty sets such that WE_r stands for the set of exit states for the role $r \in R$;
- $(WSG_r)_{r \in R} \subseteq W$ is a family of non-empty sets such that WSG_r stands for the set of stay and go states for the role $r \in R$;
- $\Theta \subseteq W \times W$ is a set of directed edges;
- $\lambda : \Theta \longrightarrow (\mathcal{L}_C \times (\mathcal{L}_{CI} \cup \mathcal{L}_E^{integer} \cup \mathcal{L}_E^{real}) \times \mathcal{L}_A)^n$ is a labelling function. Each arc is labelled with a set of triples of a constraint expression, an illocutionary scheme or timeout, and an action language expression;
- $min, Max : R \longrightarrow \mathbb{N}$ $min(r)$ and $Max(r)$ return respectively the minimum and maximum number of agents that must and can play the role $r \in R$;
- i stands for the scene information model;

satisfying the requirements below:

1. (W, Θ) is a connected graph.
2. Every scene state is reachable from the initial state and connected to a final state. Formally, $\forall w \in W, \exists w_0, \ldots, w_m \in W$ such that $w_m \in W_f$ and $\forall i \in [1 \ldots m], (w_i, w_{i+1}) \in \Theta$ and $\exists j \in [0 \ldots m]$ such that $w = w_j$.

---

[9] Some syntactic simplifications can be introduced to avoid writing many such constraints by prefixing variables with the symbol '?' to indicate that they are free and with the symbol '!' to indicate that the value of the variable has to be the same as the last bound one.

3. Each final state is an exit state for each role. $\forall r \in R, W_f \subseteq (WE_r)$
4. The initial state belongs to the access states for those roles whose minimum is greater than zero. $\forall r \in R, min(r) > 0 \Rightarrow w_o \in WA_r$
5. The maximum number of agents per role is equal to or greater than the minimum, which has to be equal to or greater than zero. $\forall r \in R, Max(r) \geq min(r) \geq 0$
6. The initial state is non-reachable once left. $\nexists w \in W$ such that $(w, w_0) \in \Theta$
7. There are no outgoing arcs from the final states. $\forall w_i \in W_f, \nexists w_j \in W$ such that $(w_i, w_j) \in \Theta$
8. The protocol is deterministic. For illocutionary schemes this means that different arcs outgoing from the same state cannot be labelled with illocutionary schemes that can be unified.
9. All the occurrences of a variable within a scene must have the same type.

2.4 Performative Structure

In the context of an institution, as in business process modelling, activities are inter-related. A performative structure is precisely the specification of these relationships:

- to capture *causal dependencies* among activities (e.g. a patient cannot undergo an operation without being previously diagnosed by a doctor);
- to define *synchronisation* mechanisms (e.g. within an exchange house, we might require the simultaneous presence of a certain number of traders to start off a negotiation scene);
- to establish *parallelism* mechanisms (e.g. in an auction house, several auctions — possibly devoted to the auctioning of heterogeneous types of goods under the rules of different auction protocols — might be run at the same time);
- to define *choice points* that allow agents leaving an activity to choose their destination (e.g. an agent attending a conference is expected to opt for just one among various simultaneous talks);
- to establish the *role flow policy* among activities, i.e. which paths (or sequences of activities) can be followed by the agents and which roles they can incarnate in each activity (e.g. in a conference centre, a speaker, after finishing his talk, is permitted to leave the conference room and go to another conference room to become a listener of another talk).

A performative structure can thus be understood as a collection of multiple concurrent activities. Agents navigate among activities constrained by the rules defining the relationships among them. Moreover, the very same agent might be participating in multiple activities at the same time, and even playing different roles in each.

From a structural point of view, performative structures may be regarded as a network of activities. A basic activity (node) is a scene specifying a conversation. A node can also be a complex activity and thus a node can be the specification of a performative structure itself. That is, performative structures can recursively contain other performative structures. In what follows, we use the word *activity* to refer to both scenes (simple activities) and performative structures (complex activities) of a performative structure. At this point we must notice that the way agents move from one activity to another depends on the type of relationship holding among the source and target activities. As mentioned above, sometimes we might be interested in forcing agents to synchronise before jumping into either new or existing activity executions, or offer choice points so that an agent can decide which target activity to enter into. In order to model the relationships listed above we use a special type of scene, the so-called *transition* scenes.[10] Each activity may be connected to multiple transitions, and in turn each transition may be connected to multiple activities. These connections are actually modelled as directed *arcs*.

Any performative structure must contain an initial and a final scene. They do not model any activity and must be regarded as the entering and exit points in the institution. Hence, any agent joining a performative structure is placed in the initial scene, from where it can move to the other activities within the performative structure until it reaches the exit scene and leaves the institution.

The arcs connecting transitions to scenes and performative structures play a fundamental role. Notice that as there might be multiple executions (or perhaps none) of a target activity, one needs to specify whether the agents following the arcs are allowed to start a new activity execution, whether they can choose a single or a subset of them to enter into, or whether they must enter all the activity executions. We will use labels on the arcs to distinguish these options.

We define a set of different types of transitions and arcs whose semantics will serve to constrain the mobility of agents among the activities of a performative structure. The differences between the diverse types of transitions that we consider are based on how they allow the agents that reach them to progress towards other scenes. Specifically we define the following types of transitions:

- **And**: They establish a synchronisation and parallelism point. Agents are forced to synchronise to enter the transition and to subsequently follow the outgoing arcs in parallel.
- **Or**: They behave in an asynchronous way at the input (agents are not required to wait for others in order to progress into the transition), and as choice points at the output (agents are permitted to select which outgoing arc — which path — to follow when leaving).

The label of each arc of the graph determines which agents depending on their role can progress through the arc. This is expressed as conjunctions and disjunctions of pairs of an agent variable and a role identifier. The role identifier determines which agents will be allowed to follow the arc depending on their role, while the agent variables are used to differentiate among agents playing the same role. A conjunction defines groups of agents that have to travel together through the arc. For instance, a label $(x : r_i)$ and $(y : r_j)$ means that this arc can be followed by pairs of

---

[10] Henceforth *transitions* for short. Although transitions are a particular class of scenes, hereafter we will be using the terms separately to distinguish transitions from non-transition scenes.

agents where one of them is playing the role r_i and the other is playing the role r_j. On the other hand, a label $(x : r_i)$ or $(y : r_j)$ means that any agent playing one of the roles r_i or r_j can progress through the arc alone.

Definition 2.11. We define the arc label language $\mathcal{L}_L$ as the language generated by the following grammar with starting symbol L:

$$L ::= \ C$$
$$\quad | \ L \textbf{ or } C$$

$$C ::= \ (x : r) \qquad\quad \text{with } x \in V_{Agents}, r \in (K_{R_I} \cup K_{R_E})$$
$$\quad | \ C \textbf{ and } (x : r) \quad \text{with } x \in V_{Agents}, r \in (K_{R_I} \cup K_{R_E})$$

The scope of agent variables is the incoming and outgoing arcs of the transition. That is, if an agent reaches a transition following an arc labelled with $(x : r_i)$ it can only leave the transition following arcs which contain the variable x in their label. We want to note that when the arc is connecting an activity to a transition, a conjunction means that agents have to leave the source activity together, and if the conjunction labels an arc from a transition to an activity, it means that agents must enter into the same activity execution(s). Furthermore, agents can change their role while moving among activities. This is specified by associating the same variable with different roles in the incoming and outgoing arcs of a transition. As in the case of scene specifications, performative structure arcs can also be labelled with a constraint and an action language expression. The constraint expression must be satisfied for agents to be allowed to progress through the arc, and the action language expression is executed once the agents have progressed throughout the arc.

Summarising, a performative structure is specified as a graph with three types of nodes, namely scenes, performative structures and transitions, connected by means of directed arcs establishing how agents can legally move between them. In this context, while scenes and performative structures determine the activities in which agents can engage, transitions capture the relationships among them. Agents within a performative structure devote their time to jointly start new activity executions, to interact with other agents in the different scenes in which they are taking part, to leave current activities to enter into other ones, and finally to abandon the performative structure. In this context, to start a scene or performative structure execution must be regarded as the instantiation of a particular node. When a performative structure is instantiated just an instance of the initial and final scenes is created. Executions of the rest of the scene and performative structures are created as agents progress through an arc of type *new*. The performative structure must guarantee that each scene and performative structure can be reached and left by agents playing their roles. Notice that in the case of performative structure nodes the roles that can enter and leave them may be different, as agents can change their role when moving between scenes. Specifically, the roles that can enter a performative structure are the roles of its initial scene, while agents can leave a performative structure with any of the roles of its final scene.

Next, we bundle all the elements introduced above to provide a formal definition of a performative structure specification.

Definition 2.12. A performative structure type is a tuple

$$ps = (W, T, s_0, s_\Omega, E, f_{E_I}, f_{E_O}, f_L, f_C, f_A, f_W, f_T, f_\mathcal{E}, C, \mu, i)$$

where

- $W \subseteq K_{state}$ is a finite, non-empty set of node identifiers, such that $W = W_S \cup W_{PS}$ where W_S stands for the scene node identifiers and W_{PS} stands for the set of performative structure node identifiers;
- $T \subseteq K_{transition}$ is a finite and non-empty set of transition identifiers;
- $s_0 \in W_S$ is the *initial* scene;
- $s_\Omega \in W_S$ is the *final* scene;
- $E \subseteq K_E$ is a finite, non-empty set of arc identifiers, such that $E = E_I \cup E_O$ where E_I is a set of arc identifiers from scenes and performative structures to transitions, and E_O is a set of arc identifiers from transitions to scenes and performative structures;
- $f_{E_I} : E_I \longrightarrow (W \times T)$ returns for each arc identifier in E_I its source scene or performative structure and its target transition;
- $f_{E_O} : E_O \longrightarrow (T \times W)$ returns for each arc identifier in E_O its source transition and the target scene or performative structure;
- $f_L : E \longrightarrow \mathcal{L}_L$ is a labelling function, returning for each arc a disjunctive normal form of pairs of agent variable and role identifier expressed in $\mathcal{L}_L$;
- $f_C : E \longrightarrow \mathcal{L}_C$ returns the contraint associated with an arc;
- $f_A : E \longrightarrow \mathcal{L}_A$: maps each arc to an action language expression in $\mathcal{L}_A$;
- $f_W : W \longrightarrow S \cup PS$ maps each element in W to a scene type or a performative structure type;
- $f_T : T \longrightarrow \{and, or\}$ maps each transition to its type;
- $f_\mathcal{E} : E_O \longrightarrow \{one, some, all, new\}$ maps each arc to its type;
- $\mu : W \longrightarrow \{0, 1\}$ sets whether a scene or performative structure can be multiply instantiated at execution time;
- i: stands for the information model of the performative structure.

We need to introduce some notation before presenting the requirements to be satisfied by the items above. (i) In order to denote the elements of the scene type of a scene node, we write the node identifier as a super index. For instance R^s denotes the set of roles of the scene type of s. (ii) By $inputRoles^{ps}$ and $outputRoles^{ps}$ we denote the sets of roles that can enter and leave a performative structure ps, respectively. Notice that agents can change their roles during their participation in a performative structure. Hence, the sets of input and output roles of a performative structure can be different. (iii) By $incomingArcs^w$ and $outgoingArcs^w$ we denote the sets of incoming and outgoing arcs of node w respectively. (iv) Let var, $roles$ and $conjunctionRoles$ be three functions over arcs, defined respectively as $var : E \longrightarrow 2^{V_{Agents}}$, $roles : E \longrightarrow 2^{Roles}$ and $conjunctionRoles : E \longrightarrow 2^{2^{Roles}}$. The functions var and $roles$ return for each arc the set of variables and the set of roles contained in the arc label respectively, while *conjunctionRoles* returns a set of sets of roles, one set for each conjunct in the label. Notice that the last function is important because agents playing the roles appearing in a conjunct must enter the target scene together. (v) *PathTo* and *PathFrom* are two functions that check node reachability. They are defined as $PathFrom : W \times W \times Roles \longrightarrow boolean$ and $PathTo : W \times W \times Roles \longrightarrow boolean$. $PathFrom(w_i, w_j, r)$ checks whether there is a path that allows agents playing role r in w_i to reach w_j; $PathTo(w_i, w_j, r)$ checks whether there is a path that allows agents playing one of the roles in w_i to reach w_j with role r.

The following requirements must be satisfied by every performative structure:

1. For each scene and for each role of the scene there is a path from the initial scene which will allow agents to reach the scene with that role. $\forall s \in W_S$ and $\forall r \in R^s, PathTo(s_0, s, r)$
2. From each scene and for each role that can be played within the scene there has to be a path that permits agents playing that role to leave the institution, i.e. there is a path to the final scene. $\forall s \in W_S$ and $\forall r \in R^s, PathFrom(s, s_\omega, r)$
3. For each performative structure node and for each role that can enter that performative structure there has to be a path from the initial scene that will allow agents to reach the performative structure with that role. $\forall ps \in W_{PS}$ and $\forall r \in inputRoles^{ps}, PathTo(s_0, ps, r)$
4. From each performative structure node and for each role that can leave that performative structure there has to be a path that permits agents playing that role to leave the institution, i.e. there is a path to the final scene.
   $\forall ps \in W_{PS}$ and $\forall r \in outputRoles^{ps}, PathFrom(ps, s_\omega, r)$
5. All and only the agent variables labelling the incoming arcs of a given transition must also appear on the outgoing arcs of the transition.
   $\forall t \in T, \bigcup_{e_i \in incomingArcs^t} var(e_i) = \bigcup_{e_j \in outgoingArcs^t} var(e_j)$
6. For every scene, the roles labelling its incoming arcs must belong to its role set.
   $\forall s \in W_S$ and $\forall e \in incomingArcs^s, roles(e) \subseteq R^s$
7. For every scene, the roles labelling its outgoing arcs must belong to its role set.
   $\forall s \in W_S$ and $\forall e \in outgoingArcs^s, roles(e) \subseteq R^s$
8. For each performative structure node the roles labelling its incoming arcs must belong to roles that can enter that performative structure.
   $\forall ps \in W_{PS}$ and $\forall e \in incomingArcs^{ps}, roles(e) \subseteq inputRoles^{ps}$
9. For each performative structure node the roles labelling its outgoing arcs must belong to roles that can leave that performative structure.
   $\forall ps \in W_{PS}$ and $\forall e \in outgoingArcs^{ps}, roles(e) \subseteq outputRoles^{ps}$
10. For every arc of type *new* the initial state of the scene must belong to the set of access states for each role labelling the arc. $\forall e \in E_O$ such that $f_{E_O}(e) = (t, s)$, if $s \in W_S$ and $f_{\mathcal{E}}(e) = $ new then $\forall r \in roles(e), w_0^s \in WA_r^s$
11. For each node, except for the initial and final scenes, there must exist an incoming arc of type *new*. $\forall w \in W, w \neq s_0, w \neq s_\Omega, \exists e \in incomingArcs^w$ such that $f_{\mathcal{E}}(e) = $ new

12. For each set of roles appearing together in a conjunction in an incoming arc of
    a scene there is a scene state which is an exit state for all of them.
    $\forall s \in W_S, \forall e \in incomingArcs(s)$ and $\forall c \in conjunctionRoles(e), \bigcap_{r \in c} WE_r^s \neq \emptyset$

2.5 Normative Rules

As described so far, the performative structure defines what participating agents
are *permitted* to do within the institution depending on their role. The performative
structure constrains the behaviour of participating agents on two levels:

- *intra-scene:* Scene protocols dictate for each agent role within a scene what can
  be said, by whom, to whom and when.
- *inter-scene:* The connections between the scenes of a performative structure
  define the possible paths that agents may follow depending on their roles. Furthermore, the constraints over output arcs impose additional restrictions on
  agents attempting to reach a target scene.

Some agent actions *within* scenes may have consequences that either limit or ex-
pand the agent's subsequent potential actions, beyond the scope of the state transition
in a scene or beyond the scope of the scene. The consequences we have identified can
take two different forms. Some actions create commitments for future actions, which
may be interpreted as obligations. Other actions may affect the paths an agent may
take through the performative structure because it may change which constraints are
satisfied. For instance, a trading agent winning a bidding round within an auction
house is obliged to pay later on for the acquired goods before leaving the institution. In order to capture these consequences, we use a special type of rules we call
normative rules.

Since we are specifying dialogical institutions, agents' actions are expressed as
a pair <illocutionary scheme ; scene where it is uttered>, or as triplets if we want
to also specify in which state the illocution is uttered. We need these components
because the same illocution could appear in more than one scene or in more than one
state. The scene gives the context in which the illocution must be interpreted and,
of course, this affects the consequences that the utterance of the illocution has. That
is to say, the same illocution may have different consequences in different scenes
because it is uttered in a different context. As we have said, some of the terms of
an illocutionary scheme are variables. The activation of a norm may depend on the
values of these variables in the uttered illocutions. Then, some conditions on the
value of these variables can be imposed. These conditions are specified as boolean
expressions over illocutionary scheme variables and a norm will not be activated if
they are not satisfied. The scope of the variables is the complete norm.

We define a predicate that will allow us to express the connection between illocutions and normative rules: $uttered(s, [w,]\varphi)$ denotes that a grounded illocution

unifying with the illocutionary scheme φ has been uttered in state w of scene s. The state is an optional parameter.

Definition 2.13. Normative rules are first-order formulas of the form
$$\left(\bigwedge_{i=1}^{m} uttered(s_i, [w_i], \varphi_i) \land \bigwedge_{j=0}^{n} e_j \right) \rightarrow \left(\bigwedge_{k=1}^{p} uttered(s_k, [w_k], \varphi_k) \land \bigwedge_{l=0}^{q} e_l \right)$$
where the e symbols represent boolean expressions over variables from the different illocution schemata φ.

Notice that $\varphi_1, \ldots, \varphi_p$ can be regarded as *obligations* that agents acquire whenever the antecedent of the normative rule is satisfied. Therefore, agents must utter grounded illocutions matching these illocution schemas and satisfying $e_1, \ldots, e_p$, in the corresponding scenes in order to avoid violation of the normative rule.

Clearly, an external agent might not fulfil its obligations. As agents are autonomous and the institution accepts agents developed by other people, those agents cannot be forced to utter particular illocutions. From this it follows that institutions cannot force agents to fulfil their obligations. However, the institution knows the obligations that each agent has acquired and can thus detect when an agent does not fulfil its obligations and hence violates the normative rules. Moreover, the institution can restrict the actions that an agent can carry out while it has not fulfilled some or all of its obligations.

2.6 Electronic Institution

Finally, an electronic institution specification consists of a dialogical framework, a performative structure and a set of normative rules.

Definition 2.14. An electronic institution is defined as a tuple

$$ei = (df, ps, N)$$

where

- df stands for a dialogical framework;
- ps stands for a performative structure;
- N stands for a set of normative rules.

2.7 Operations and Operational Semantics

The EI metamodel postulates, on one hand, the means to specify an electronic institution, and that is what we covered in some detail in the past five sections. On the other hand, the EI model also postulates the way an electronic institution is intended to work and that is what we cover in this section in a *superficial* way.

2.7.1 Operations

One distinction to keep in mind is that in order for an electronic institution (as defined in Def. 2.14) to work, there are those actions that agents take and there are those actions that the electronic institution as the social environment itself needs to support. Thus,

- an agent that is in the environment of an electronic institution can only speak, keep quiet or move between scenes (in fact move from scenes to transitions and from transitions to scenes and change roles within a transition); however,
- the environment needs to become active and terminate, scenes need to be created and closed, agents need to be let in and out of the environment, as well as allowed to move from one scene to another through a transition.

In other words, the electronic institution — as an *institutional infrastructure* that creates a social environment where agents interact — needs to support actions that are called either by an agent or by the infrastructure itself. And these actions have effects on constructs of the environment, namely, on a scene, a transition or the environment as a whole.

Table 2.7.1 lists all the operations involved in the EI metamodel (and are presented and discussed in detail in [13]).[11]

Operation	Called by	Effect on
Speak	Agent	scene
RequestAccess	Agent	electronic institution
JoinInstitution	Agent	electronic institution, scene
LeaveInstitution	Agent	electronic institution, scene
SelectNewTargets	Agent	transition
RemoveOldTargets	Agent	transition
StartElectronicInstitution	Infrastructure	electronic institution
CreateSceneInstance	Infrastructure	scene
CloseSceneInstance	Infrastructure	scene
EnableAgentsToLeaveOrTransition	Infrastructure	transition
EnableAgentsToLeaveAndTransition	Infrastructure	transition
MovingFromSceneInstanceToTransitionInstance	Infrastructure	scene, transition
MoveAgentFromTransitionToScene	Infrastructure	scene, transition
RemoveClosedInstances	Infrastructure	electronic institution
Timeout	Infrastructure	scene

Table 2.1 Electronic institution operations (the last column indicates the constructs that the operation updates)

---

[11] Recall that transitions — from the perspective of the environment — are just a type of scene, and that — as far as enabling agent actions — a transition is actually a very special type of scene where several constraints may need to be enforced by the system.

2.7.2 Operational Semantics

The operational semantics of the EI metamodel are quite straightforward — and also rigorously defined in [13].

An agent can speak by uttering an illocution, which, if it is grounded and if it satisfies the conditions that are active in the scene and institutional states at that very moment, will have the effects on the scene state (and on the institutional state) that the action language prescribes for that action in the corresponding scene specification. Likewise, the agent may move from a scene to a transition with a proper utterance, it may change roles if the prevailing constraints allow that change, and with a proper utterance may then leave that transition and enter an intended scene as long as the transition and goal scene constraints are satisfied.

The operational semantics of the stay and go scene states is a bit more involved inasmuch as an agent that chooses to stay and go would unfold as two "agent processes" provided that the corresponding circumstances allow it. An agent may unfold in as many "agent processes" as needed, keeping in mind that any such process stands for the same agent and each process is bound by the constraints that affect that process in whatever action it intends to accomplish. The key point is that in particular, the conditions and the effects that affect a particular agent process *involve the values of the current state of the agent as a whole*. For example, an agent may be simultaneously active in several auctions but if it wins a bid in any of these, its credit will be altered in that instant and, consequently, this will affect its possibilities of winning a bid in any other ongoing auction.

The operational semantics of joining a transition are related to the labelling of (transition) incoming arcs, to allow the transit of an agent with a given role out of a given scene. However, the semantics of the operation of leaving a transition involve not only the roles that may be played in the goal scene — which must be labelled in the corresponding outgoing arc of the transition — but also the type of transition (AND, OR), the compatibility of roles that an agent wishes to switch between, the existence or the need to create a new instance of the goal scene, and the constraints that apply to that goal scene (e.g. the prevailing number of agents that are active in that scene).

Timeouts assume that the institutional infrastructure may keep track of time. The other operations called by the infrastructure correspond to the obvious intuitions (for details see [13], again).

2.8 Implementation

Implementation of electronic institutions requires a platform, and this involves a computational architecture that is embedded in the technological environment where the EI is to run, and a set of tools that translate the conceptual framework into computable features, namely a specification language and tools to test, run and monitor electronic institutions. In this section we give a rough description of one such platform, the *Electronic Institutions Development Environment*, EIDE [16, 3].

EIDE involves:

- a particular computational architecture (the EIDE architecture),
- a particular specification language, ISLANDER [15], and
- a particular set of tools, which includes: (i) AMELI, middleware to run an ISLANDER–compatible specification of an electronic institution on the EIDE architecture, (ii) tools to monitor a running institution, (iii) SIMDEI, a monitoring add-on to perform simple simulations on an electronic institution, and (iv) aBUILDER to deploy "agent skeletons" that comply with the ISLANDER specification of an institution.

As we shall discuss in the next section, other computational architectures exist, some compatible with ISLANDER, others not, and there are other specification languages that are compatible with EI — because they extend or restrict in some way the expressiveness of the EI metamodel — and these, in turn, may run on an adapted EIDE architecture or on completely different ones.

2.8.1 The EIDE Architecture

Any institutionally relevant action in any institution, conventional or electronic, needs to take place within the institutional environment (as we stipulated in Sect.1.2.1, page 8). Accordingly, the EIDE architecture creates an environment for the electronic institution where all agent interactions take place. What is distinctive about the EIDE architecture is that this is a restricted environment where every interaction among agents and between agents and the institution is mediated by the electronic infrastructure itself [14]. As Figure 2.6 describes, each actual agent — be it external or internal — has its own *governor*, which is an "infrastructure agent" provided and controlled by the institution, and it is through this *governor* that all contact between agent and institution takes place.[12]

Figure 2.6 also shows how each governor is in touch with all those scenes and transitions where its agent is active at any particular time, and shows also that every transition and every scene has other infrastructure agents — transition and scene managers — that communicate with governors and with another infrastructure agent, the institution manager.

There are two features of this setup worth remarking. First, that the process of keeping a *unique updated state of the institution* is the key task that has to be guaranteed by the institutional infrastructure, and second, that it is a complex task that

---

[12] In terms of the description we made in Sect. 2.1, external and internal agents exist in $\mathcal{W}$ and each has a corresponding *governor* in $\mathcal{T}$, whose behaviour is determined by the external or internal agent but has to comply with the institutional conventions specified in $\mathcal{I}$.

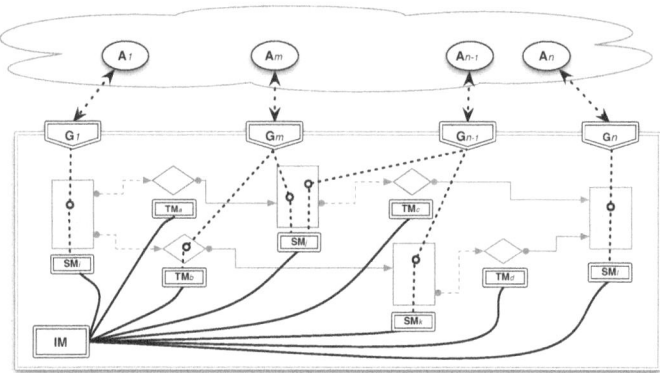

Fig. 2.6 A computational architecture for electronic institutions (after [13]). Participating agents (A) communicate with (infrastructure) governors (G), which in turn coordinate with other infrastructure managers for each scene (SM) and each transition (TM), and with the institution manager (IM)

involves a sophisticated control of information flows that originate in the different scenes and transitions that are active at any given moment.[13]

Governors deserve a special commentary. A governor constitutes the *de facto* interface between its agent and the institution. Note that the governor has no access to its agent decision model, it is a mere interface. Thus, an agent will receive only that information that its governor provides, and the only information that will ever get into the institution is that which the governor transmits. This can be accomplished because the governor has a copy of the specification of the institution and it also keeps an updated version of the part of the institutional state that affects its agent. Consequently, as an input interface, when its agent attempts to "speak" or "move", the governor admits into the institution only those utterances that are actual grounded illocutions that satisfy the constraints of the scene state or the transition where the agent attempts to speak or move. Then, these admitted illocutions can change the state of the institution the way they are intended to. In the other direction, the governor communicates to its agent only those messages that this agent is entitled to hear according to the electronic institution specification. A few comments:

1. Governors are a feature of the EIDE architecture and *not a feature* of the EI metamodel.

---

[13] In practice, the *institution manager* is in charge of the admission and termination of agents (thus, it creates governors, communicates with them and terminates them) and of creating new scene and transition instances and terminating them (thus, it creates, communicates with and terminates scene and transition managers). *Scene and transition managers* control access and departure of agents and the execution of scenes and transitions. For this purpose they communicate with those managers that are connected to them and with the governors of agents that are active in their scenes or transitions.

2. This idea of a "governed interface" provides a powerful governance mechanism because the governor in fact enforces *all regimented constraints*. Nonregimented constraints need to be specified and handled through internal agents.
3. In terms of the FIPA reference model [17], the institutional infrastructure occupies a level above the agent communication layer. Thus, the governor receives and transmits messages from and to an agent communication support environment (such as JADE). The EIDE platform assumes only that this communication layer is FIPA-compliant. An additional layer would be needed to create a human interface.[14]

2.8.2 Tools

ISLANDER [15] is a specification language that models electronic institutions according to the EI metamodel described in the previous sections. It is a graphical software tool — illustrated in the previous chapter (see Figures 1.3 and 1.4) — with its own versions of expressions and communication, constraint and action languages, as well as graphical representation of EI constructs such as role relationships, scenes and performative structures. It also provides automated verification of syntactic properties of the specification such as the constraints on scenes (Def. 2.10) and performative structure (Def. 2.12). Thus, it allows validation of language and structural integrity as well as liveliness of the specification. ISLANDER generates an XML specification file that serves as input for the AMELI middleware to enact it and aBUILDER to generate agent skeletons.

Based on the ISLANDER specification, AMELI enacts a working environment that is an instantiation of the EIDE architecture. This AMELI-enacted environment activates the infrastructure agents as needed, controls the activation and cancellation of scenes and transition instances, and enables the access of agents and their actions (speak or move). Thus AMELI takes care of the correct evolution of scenes and the movements of agents through scenes and transitions.

The EIDE platform includes a *monitoring tool* that allows the inspection of the evolution of an active electronic institution. It can display the discrete events that happen during an enactment and may focus on the activity of an agent, or on a scene or on transitions. This monitoring tool is accessible to SIMDEI. While ISLANDER provides static verification of properties, the SIMDEI tool allows some dynamic verification through discrete event simulations of AMELI. EIDE provides a simulation bridge to connect SIMDEI with external simulators.

Finally, EIDE includes aBUILDER, a tool to deploy agent skeletons (code) that can be used on SIMDEI simulations or in an enactment of an ISLANDER specification. These skeletons are automatically generated from an ISLANDER specification file and are capable of navigating the whole institution but are "empty" in the sense

---

[14] See [4] for a more general discussion about this layered notion of environment.

that they require a decision model to house the contents of a potential illocution and decide whether or not to utter it when they have a chance.

2.9 Other EI Metamodels and Platforms

Over the twenty years since the first implementations, there are have been several proposals of alternative metamodels and platforms. Without attempting a full survey of these, we may identify three main strategies that have guided those efforts.

The first line is related to the possibility of a more explicit use of normative concepts. The point is to have normative expressiveness and to deal with non-regimented governance in a less ad hoc way. It should be noted that the EI metamodel contemplates the possibility of normative language [13] and in Sect. 2.5 we mentioned normative rules as a crude way of dealing explicitly with commitment propagation. Moreover, constitutive norms are specified as part of the dialogical framework, while procedural and functional norms are stated in scene specifications and transition constraints. But truer normative forms have been developed. For instance, there is a proposal [20] to use a normative language to express institutional conventions and to attach inference engines to the implementation architecture. Another proposed extension [22, 19] is to deal explicitly with norms and normative conflicts through the use of a "normative structure" that handles norms and propagation of normative consequences between scenes. Garcia-Camino's Ph.D. thesis has a good discussion of these issues [21].

A second line of variants has to do with the evolution of an electronic institution. The EI framework has no primitive operations that allow agents to (a) add a scene to a performative structure, (b) change the specification of a scene, nor (c) create new agents. These limitations can be circumvented to some extent. Limitation (a) may be circumvented by grafting instances of performative substructures or scenes that have been previously specified when appropriate conditions are met. Limitation (b) may be overcome with "transition functions" that, for example, define scene parameters that are changed at run-time, either by including the conventions that change the specification as part of the scene specification or by programming them in internal agents. Finally, new *internal* agents may be created by an internal agent that invokes a service that spawns new agents. There have been proposals for other forms of autonomous adaptation, for instance [8, 10], to make an electronic institution evolve or change.

A third line of developments have to do with "ergonomic concerns". EI/EIDE has been criticised for the complexity entailed by its expressiveness. Practice has shown that for many applications the complete framework apparatus is excessive. One can trim the scene specification language, the representation of transitions and not allowing any agent to be active simultaneously in more than one scene and still have a powerful coordination framework as demonstrated in the following chapters with SIMPLE and PEERFLOW. The consequence is a simpler way of specifying an electronic institution and a lighter-weight technological platform.

References

1. Alberts, L.K.: YMIR: An ontology for engineering design. Ph.D. thesis, University of Twente (1993)
2. Almajano, P., Mayas, E., Rodriguez, I., Lopez-Sanchez, M., Puig, A.: Structuring interactions in a hybrid virtual environment - infrastructure & usability. In: Proceedings of the International Conference on Computer Graphics Theory and Applications and International Conference on Information Visualization Theory and Applications (VISIGRAPP 2013), pp. 288–297 (2013). DOI 10.5220/0004215802880297
3. Arcos, J.L., Esteva, M., Noriega, P., Rodríguez-Aguilar, J.A., Sierra, C.: Engineering open environments with electronic institutions. Engineering Applications of Artificial Intelligence **18**(2), 191–204 (2005). DOI http://dx.doi.org/10.1016/j.engappai.2004.11.019. URL http://www.sciencedirect.com/science/article/pii/S0952197604001848
4. Argente, E., Boissier, O., Carrascosa, C., Fornara, N., McBurney, P., Noriega, P., Ricci, A., Sabater-Mir, J., Schumacher, M.I., Tampitsikas, C., Taveter, K., Vizzari, G., Vouros, G.A.: The role of the environment in agreement technologies. Artificial Intelligence Review **39**(1), 21–38 (2013)
5. Becht, M., Gurzki, T., Klarmann, J., Muscholl, M.: ROPE: Role oriented programming environment for multiagent systems. In: Proceedings of the Fourth IFCIS Conference on Cooperative Information Systems (CoopIs'99). IEEE, Los Alamitos (1999)
6. Belakhdar, O., Ayel, J.: Modelling approach and tool for designing protocols for automated negotiation in multi-agent systems. In: W. van de Velde, J.W. Perram (eds.) Agents Breaking Away, no. 1038 in Lecture Notes in Artificial Intelligence, pp. 100–115. Springer, Berlin (1996)
7. Bogdanovych, A., Simoff, S.: Establishing social order in 3d virtual worlds with virtual institutions. In: A. Rea (ed.) Security in Virtual Worlds, 3D Webs, and Immersive Environments: Models for Development, Interaction, and Management, pp. 140–169. IGI Global, Hershey (2011)
8. Bou, E., Lopez-Sanchez, M., Rodríguez-Aguilar, J.A.: Adaptation of autonomic electronic institutions through norms and institutional agents. In: G. O'Hare, A. Ricci, M. O'Grady, O. Dikenelli (eds.) Engineering Societies in the Agents World VII, *Lecture Notes in Computer Science*, vol. 4457, pp. 300–319. Springer, Berlin (2007)
9. Buhr, R.J.A., Elammari, M., Gray, T., Mankowski, S., Pinard, D.: Understanding and defining the behaviour of systems of agents with use case maps. In: Proceedings of the Second International Conference on the Practical Application of Intelligent Agents and Multi-Agent Technology (PAAM'97) (1997)
10. Campos, J., Lopez-Sanchez, M., Esteva, M.: Using a two-level multi-agent system architecture. In: M. De Vos, N. Fornara, J.V. Pitt, G. Vouros (eds.) Coordination, Organization, Institutions and Norms in Agent Systems VI (COIN 2010), *Lecture Notes in Computer Science*, vol. 6541. Springer, Berlin (2011)
11. Castelfranchi, C.: InMind and OutMind Societal Order Cognition and Self-Organization: The role of MAS. Invited talk for the IFAAMAS "Influential Paper Award". AAMAS 2013. Saint Paul, Minn. US (2013)
12. Christiaanse, R., Ghose, A., Noriega, P., Singh, M.P.: Characterizing artificial socio-cognitive technical systems. In: A. Herzig, E. Lorini (eds.) Proceedings of the European Conference on Social Intelligence (ECSI-2014), Barcelona, Spain, November 3-5, 2014., *CEUR Workshop Proceedings*, vol. 1283, pp. 336–346. CEUR-WS.org (2014). URL http://ceur-ws.org/Vol-1283/paper_49.pdf
13. d'Inverno, M., Luck, M., Noriega, P., Rodríguez-Aguilar, J.A., Sierra, C.: Communicating open systems. Artificial Intelligence **186**(0), 38–94 (2012). DOI 10.1016/j.artint.2012.03.004. URL http://www.sciencedirect.com/science/article/pii/S0004370212000252
14. Esteva, M.: Electronic Institutions: from specification to development. Ph.D. thesis Universitat Politècnica de Catalunya (UPC), 2003. No. 19 in IIIA Monograph Series. IIIA (2003)

15. Esteva, M., de la Cruz, D., Sierra, C.: ISLANDER: an electronic institutions editor. In: Proceedings of the First International Joint Conference on Autonomous Agents and Multiagent systems (AAMAS '02), pp. 1045–1052. ACM Press, New York (2002)
16. Esteva, M., Rodríguez-Aguilar, J.A., Arcos, J.L., Sierra, C., Noriega, P., Rosell, B.: Electronic institutions development environment. In: Proceedings of the 7th International Joint Conference on Autonomous Agents and Multiagent Systems (AAMAS '08), pp. 1657–1658. International Foundation for Autonomous Agents and Multiagent Systems, ACM Press, New York (2008)
17. FIPA: FIPA 97 specification version 2.0 part 2. Tech. rep., FIPA - Foundation for Intelligent Physical Agents (1998)
18. Fornara, N., Cardoso, H.L., Noriega, P., Oliveira, E., Tampitsikas, C., Schumacher, M.I.: Modelling agent institutions. In: S. Ossowski (ed.) Agreement Technologies, *Law, Governance and Technology Series*, vol. 8, chap. 18, pp. 277–307. Springer, Berlin (2013)
19. Gaertner, D., Garcia-Camino, A., Noriega, P., Rodríguez-Aguilar, J.A., Vasconcelos, W.W.: Distributed norm management in regulated multi-agent systems. In: Proceedings of the 6th International Joint Conference on Autonomous Agents and Multiagent Systems (AAMAS '07), pp. 624–631. ACM Press, New York (2007)
20. Garca-Camino, A., Noriega, P., Rodríguez-Aguilar, J.A.: Implementing norms in electronic institutions. In: M. Pechoucek, D. Steiner, S. Thompson (eds.) Proceedings of the 4th International Joint Conference on Autonomous Agents and Multiagent Systems (AAMAS '05), pp. 667–673. ACM Press, New York (2005)
21. Garcia-Camino, A.: Normative Regulation of Open Multi-Agent Systems. Ph.D. thesis Universitat Autònoma de Barcelona, 2010. No. 35 in IIIA Monograph Series. IIIA (2011)
22. Garcia-Camino, A., Rodríguez-Aguilar, J.A., Vasconcelos, W.: A distributed architecture for norm management in multi-agent systems. In: J. Sichman, J. Padget, S. Ossowski, P. Noriega (eds.) Coordination, Organization, Institutions and Norms in Agent Systems III, *Lecture Notes in Computer Science*, vol. 4870, pp. 275–286. Springer, Berlin (2008)
23. Jacobson, I., Christerson, M., Jonsson, P., Övergaard, G.: Object-Oriented Software Engineering - A Use Case Driven Approach. Addison-Wesley, Boston (1996)
24. Jones, A., Sergot, M.: A formal characterization of institutionalized power. Logic Journal of the IGPL **4**(3), 427–446 (1996)
25. Jones, A.J.I., Artikis, A., Pitt, J.: The design of intelligent socio-technical systems. Artif. Intell. Rev. **39**(1), 5–20 (2013)
26. Kendall, E.A.: Agent roles and role models: New abstractions for intelligent agent system analysis and design. In: Proceedings of Intelligent Agents for Information and Process Management (AIP'98) (1998)
27. Noriega, P., Padget, J., Verhagen, H., d'Inverno, M.: A manifesto for conscientious design of hybrid online social systems. In: COINECAI2016, pp. 1–16. IEEE, Los Alamitos (2011)
28. Noriega, P., Padget, J., Verhagen, H., d'Inverno, M.: Towards a framework for socio-cognitive technical systems. In: A. Ghose, N. Oren, P. Telang, J. Thangarajah (eds.) Coordination, Organizations, Institutions, and Norms in Agent Systems X, *Lecture Notes in Computer Science*, vol. 9372, pp. 164–181. Springer, Berlin (2015)
29. Osman, N., Sierra, C., Sabater-Mir, J., Wakeling, J.R., Simon, J., Origgi, G., Casati, R.: LiquidPublications and its technical and legal challenges. In: D. Bourcier, P. Casanovas, M. Dulong de Rosnay, C. Maracke (eds.) Intelligent Multimedia: Managing Creative Works in a Digital World, vol. 8, pp. 321–336. European Press Academic Publishing, Florence (2010)
30. Riehle, D., Gross, T.: Role model based framework design and integration. SIGPLAN Not. **33**(10), 117–133 (1998)
31. Robles, A., Noriega, P., Cantú, F.: An agent oriented hotel information system. In: K.S. Decker, J.S. Sichman, C. Sierra, C. Castelfranchi (eds.) Proceedings of the 8th International Confonference on Autonomous Agents and Multiagent Systems (AAMAS '09), pp. 1415–1416. International Foundation for Autonomous Agents and Multiagent Systems, ACM Press, New York (2009)
32. Scott, W.R.: Organizations: Rational, Natural, and Open Systems. Prentice Hall, Englewood Cliffs (1992)

33. Searle, J.R.: The Construction of Social Reality. Free Press, New York (1995)
34. Trist, E.: The evolution of socio-technical systems. Occasional paper, Ontario Ministry of Labour **2** (1981)
35. Wooldridge, M., Jennings, N.R., Kinny, D.: A methodology for agent-oriented analysis and design. In: Proceedings of the Third International Conference on Autonomous Agents (AGENTS'99). ACM Press, New York (1999)

Part II
Applications of Electronic Institutions

Chapter 3
The uHelp Application

Nardine Osman, Bruno Rosell, Andrew Koster, Marco Schorlemmer, Carles Sierra and Jordi Sabater-Mir

When people need help with day-to-day tasks they turn to family, friends or neighbours to help them out. Finding someone to help out can be a stressful waste of time. Despite an increasingly networked world, technology falls short in combating such daily irritations. uHelp provides a platform for building a community of helpful people and supports them in finding help for day-to-day tasks. uHelp uses electronic institutions to coordinate interactions between individuals, and it relies on a trio of techniques — semantic similarity, a trust model and a flooding algorithm — to help efficiently find the most trusted volunteers for a given task request. This chapter provides an overview of the uHelp application, describes its underlying electronic institution, and presents a brief introduction to the integrated technologies that allow uHelp to effectively find the most suitable volunteers.

Nardine Osman
IIIA-CSIC, Barcelona, e-mail: nardine@iiia.csic.es

Bruno Rosell
IIIA-CSIC, Barcelona, e-mail: rosell@iiia.csic.es

Andrew Koster
eDreams Odigeo, Barcelona, e-mail: koster.andrew@gmail.com

Marco Schorlemmer
IIIA-CSIC, Barcelona, e-mail: marco@iiia.csic.es

Carles Sierra
IIIA-CSIC, Barcelona, e-mail: sierra@iiia.csic.es

Jordi Sabater-Mir
IIIA-CSIC, Barcelona, e-mail: jsabater@iiia.csic.es

© Springer Nature Switzerland AG 2024
N. Osman (ed.), *Electronic Institutions*, Artificial Intelligence: Foundations, Theory, and Algorithms, https://doi.org/10.1007/978-3-319-65605-2_3

3.1 Objectives

In this chapter, we present uHelp, a software application that provides a fully distributed platform for building and maintaining a local community of people helping each other with their day-to-day tasks, thus supporting the balancing of societal needs, which contributes to community well-being. For instance, the community could be a group of neighbours, friends or close family who turn to each other for help with performing day-to-day tasks. As a prototype scenario, we started by focusing on a community of parents who need help with dropping off their children at school, picking them up from school or babysitting their children. According to [1], being late to pick up the children is a major stress factor for working parents. From one of the interviews they conducted, they quote a mother: "the worst time is the afternoons, and trying to finish off work to leave on time to collect my son from the nursery." A follow-up study [13] indicates that one of the most severe problems that working parents encounter is coping with unexpected scheduling issues.

In uHelp, picking up children is one of the main tasks. However, we have also expanded the domain and designed a number of other tasks, including a generic task, to allow people to ask for help with their day-to-day activities. Additionally, we emphasise that there are many other communities in which similar technologies can be used. For instance, sport clubs where people try to find practice partners, or a group of commuters organising carpools.

The uHelp application essentially provides a platform to build and support a community to which members can turn to find help with day-to-day tasks [9, 8]. The uHelp application starts from the pre-existing social relationships between users. We thus assume the existence of a social network (identities and connections may be imported from various other social-networking applications, such as Facebook, or constructed automatically by accessing the address book of people's telephones when they join uHelp). The social network is used specifically to allow users to ask for help with picking up or taking care of their children. Three key technologies are then used to allow one to find help efficiently: a flooding algorithm, a computational trust model and a semantic similarity model.

When a member of the community needs help with a certain task, the *flooding algorithm* propagates this request, starting with that member's direct neighbours in the graph and flooding the request out from there until a satisfactory volunteer for the task is found. The flooding algorithm relies on the community members' trust evaluations of one another to decide whether to forward the request or not, as requests are only forwarded to trusted members. Trust evaluations are calculated automatically by the *computational trust model* based on requesters' evaluations of the past performance of the volunteer, with respect to the specific task requested. This is made possible as uHelp allows the requester to evaluate a volunteer's performance after they complete their tasks. The trust model, in turn, relies on the *semantics of the task* to find similar tasks, when little to no feedback is available for a task. The computed semantic similarity measures essentially help the trust model use evaluations of a volunteer performing tasks in the past to estimate his performance at a similar future task.

The remainder of this chapter is divided as follows. Section 3.2 presents the implementation of the uHelp app; Section 3.3 presents the electronic institution specification underlying the uHelp app; and Section 3.4 describes the integrated technologies that help to find volunteers efficiently.

3.2 Illustration

In this section we present the user interface of the uHelp application. The implementation was carried out using the Ionic framework,[1] which allows the development of the application as a web application that is automatically translated as well into iOS and Android applications. This allowed us to build a cross-platform application. We also made use of the Apache Cordova platform[2] for developing plugins that map device hardware to plugins in JavaScript, which could then be used on the Ionic framework.

In order to use uHelp it is necessary for the user to configure her data upon downloading the application. She must first either login via Facebook or register with the system and import her connections from her list of contacts, to create her social network. Once this is done, the user can start to use uHelp to request help, and be contacted by others in the community who need help in return.

The uHelp application's usability is designed around four main views: *1) the "Help" view*, where users may request help; *2) the "Requests" view*, where users can track and manage requests for help (issued by them or others); *3) the "Community" view*, where users can track and manage their community or social network; and *4) the "Settings" view*, where users can change their user-specific settings, as illustrated shortly. In what follows, we go over each of the uHelp application views in a bit more detail.

1. **Asking for help.** Asking for help is done by pressing the "Help!" button (see Figures 1(a) and 1(c)). However, the user must first choose the action that she requires help with from the list of tasks. The elements of this list, and the options given, correspond to elements in the predefined ontology. We note that different actions will require different parameters. To keep the application user-friendly, once an action is chosen from the list (Figure 1(b)), the appropriate fields will appear where the user can fill in the details of the requested action (Figure 1(c)). For example, if the action is "Care for a relative", then the relative field appears, along with the pickup and drop-off location and time, but if the action is "Get me something", then the object field appears, along with the pickup and drop-off location and time.

   Only after filling in the required parameters can the user proceed to pressing the "Help!" button. This button then triggers a flooding algorithm that spreads the help request in order to find volunteers, as described in Section 3.4.1. When a

---

[1] https://ionicframework.com

[2] https://cordova.apache.org

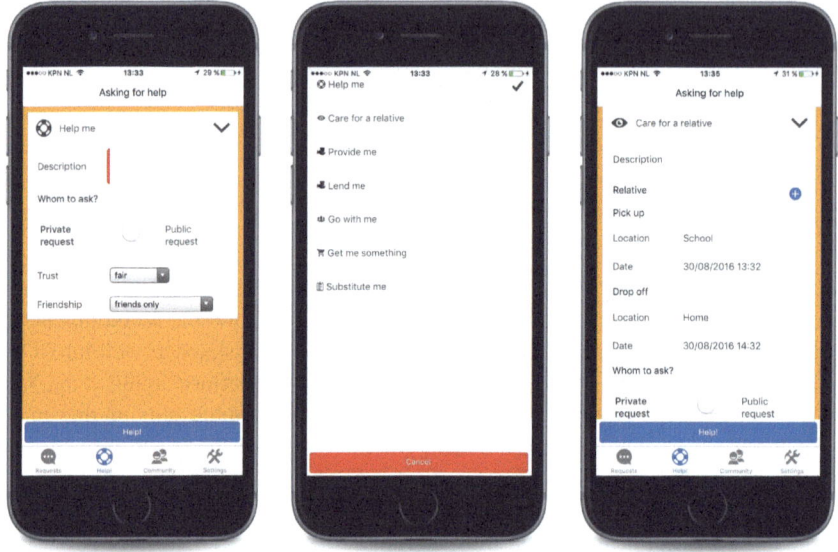

(a) The general task   (b) List of predefined tasks   (c) "Care for a relative" task

Fig. 3.1 Various screens from the "Help" view

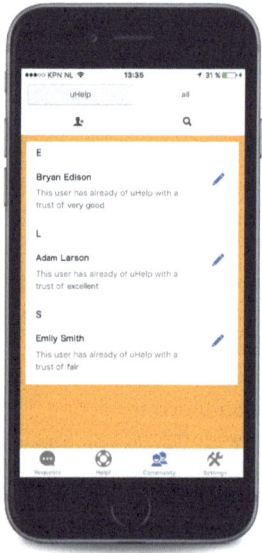

Fig. 3.2 The "Requests" view   **Fig. 3.3** The "Community" view

State	Colour	Description
looking for volunteers	yellow	Waiting...
pending assignment 1	green	Choose volunteer!
pending assignment 2	red	Choose volunteer now!
assigned	yellow	Help on its way
completed	green	Please rate!
rated	grey	Rated
cancelled	grey	
expired	grey	

Table 3.1 The state of a requester's task and its representation

State	Colour	Description
unanswered	green	Can you help?
declined	grey	I cannot help :(
accepted	yellow	Waiting to be selected ...
selected	red	Do it!
not selected	grey	Help no longer needed, thanks!
completed	grey	Completed
cancelled	grey	
expired	grey	

Table 3.2 The state of a requestee's task and its representation

request is received by another user, she receives a notification and is asked to respond.

2. **Tracking requests.** Users can see requests for help (sent by or to them) along with their state in the "Requests" view. In Figure 3.2, green text bubbles (to the right) represent one's own requests, while grey ones (to the left) represent requests received. Again, the presentation of each item in the list is dependent on the semantics of the action. In the case of the "Care for a relative" action, the name of the child and the pickup location will be displayed. Additionally, the "state" parameter is presented to help the user track the progress of requests, such as 'waiting for volunteers', 'help on its way' or 'completed'.

The state for the requestee is different from that for the requester. Tables 3.1 and 3.2 present the different states for each, along with how this state is visualised to the user. The visualisation is executed by assigning a colour along with a piece of text describing that state (see Figure 3.2). We use colours to categorise states into four categories: (1) those that need to be acted upon, which get the colour green; (2) those that need to be acted upon *immediately* (for example, when one needs to urgently select a volunteer as the deadline to select volunteers is approaching), which get the colour red; (3) those that require no action at the moment, as others are expected to be acting upon those tasks, which get the colour yellow; and (4) those that are considered closed, which get the colour grey.

The requester's actions that are permitted in this view are: *1*) cancel requests, where the cancel button should be active from the moment the request is made until a predefined amount of time preceding the task's deadline; *2*) chat with volunteers, where the requester can have private chats and even calls with the volunteers if needed; *3*) select volunteer, where the requester can select her volunteer from the list of available volunteers; and *4*) rate requests, where the rating bar should only be activated if the task has been completed or the deadline has passed.

The volunteer's actions that are permitted in this view are: *1*) accept, where the volunteer accepts to volunteer for some task; *2*) decline, where the volunteer declines to volunteer for some task; *3*) cancel, where a volunteer can cancel after accepting, and the cancel button should be active from the moment the task is accepted until a predefined amount of time preceding the task's deadline; *4*) chat with requester, where the volunteer can have a private chat or even call with the requester if needed; *5*) complete task, where the volunteer can mark her task as completed before the deadline has passed.

3. **Community.** In the "Community" view (Figure 3.3), a user can see the list of uHelp community members in her social network and can view (or even manually adjust, if needed) the trust she places in each of those members.

4. **Settings.** The final view for the user is the "Settings" view. This is where the user specifies all the information needed by the uHelp platform. The main menu (Figure 4(a)) gives a good overview of what can be changed. For example, the user can edit her profile, adding information required for the various tasks. In this case, we show how information about her children and the locations where

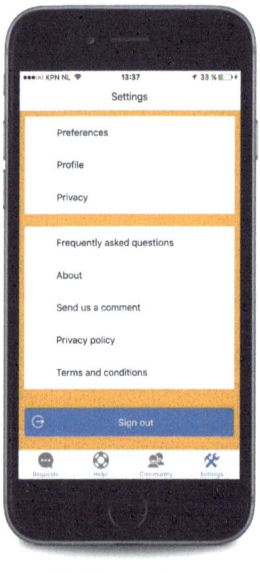
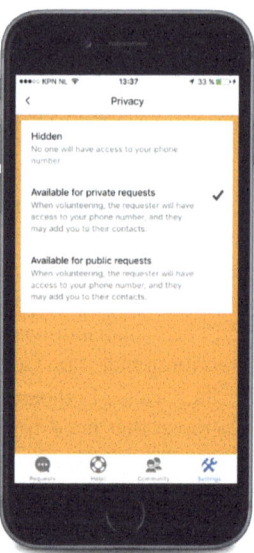

(a) "Settings" view   (b) Children & addresses

Fig. 3.4 Various screens from the "Settings" view

they need to be picked up or brought to may be edited. Additionally the user should be able to select her privacy setting, for instance, whether requesters can contact her by phone or not (Figure 4(b)).

3.3 The EI Specification

In this section we show the underlying electronic institution (EI) specification [2] of the uHelp application.

3.3.1 Dialogical Framework

In the EI model of uHelp, we can distinguish between three main roles:

- User, which is the main agent representing the human user, and is the agent that may ask for help, or receive and propagate help requests;
- Requester, which is the role that the agent representing the human user takes when dealing with a specific request that has been made by that human user; and
- Requestee, which is is the role that the agent representing the human user takes when dealing with a specific request that has been made of that human user.

Basically, each human user agent will be represented by the *User* agent. However, every time a request for help is made, the agent will "stay and go" to another scene where it plays the role of *Requester* or *Requestee* to deal with that request.

Additionally, we have one software agent:

- TrustAgent, which is the agent that updates one's trust in another and has access to the database that keeps track of users' ratings of each other along with the help requests that users receive (as required by the flooding algorithm, see Lines 4–10 of Algorithm 1).

3.3.2 Performative Structure

The workflow of the uHelp application can be divided into three stages: asking for help, flooding a help request, and dealing with help requests. These three stages can naturally be represented in an EI specification as three separate scenes, "Help" scene, "Flooding" scene and "Requests" scene. The first scene will have exactly one instance, where everyone is allowed to ask for help. However, the second scene will have as many instances as there are help requests issued. All user agents can be in more than one scene at the same time (and even in more than one instance of the second and third scenes, as more than one help request may be flooded at a time). The user agents can even play different roles in different scenes. For instance, the

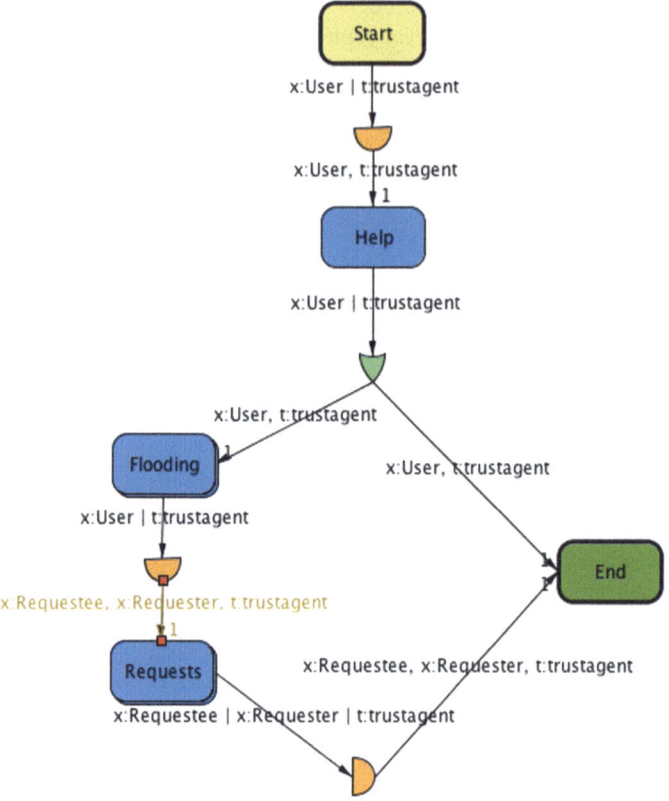

Fig. 3.5 The performative structure of the uHelp application

agent may be a *Requester* in one scene and a *Requestee* in another. The idea is that every human user will be represented by a *User* agent in the "Help" scene (asking for help). Then every time a help request is issued in that scene, a new instance of the "Flooding" scene is created for that request, which everyone moves to by performing a "stay and go". In the "Flooding" scene, the user agents (with the help of the trust agents) will propagate a help request as illustrated by the flooding algorithm of Section 3.4.1. However, the first time a request is made, the requester creates a new instance of the "Requests" scene for this request. Then, every user agent that receives this request in the 'Flooding" scene will move to the "Requests" scene instance by performing a "stay and go". In this new scene, the agent issuing the request will play the role of *Requester* and the agents receiving that request will play the role of *Requestee*. The performative structure of the institution is presented in Figure 3.5. Next, we present the three scenes in more detail.

1. **Help scene**: In this scene, the users (represented by their *User* agents) can issue help requests when needed. In addition to sending those messages, upon

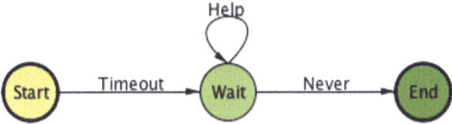

Fig. 3.6 The uHelp helping protocol

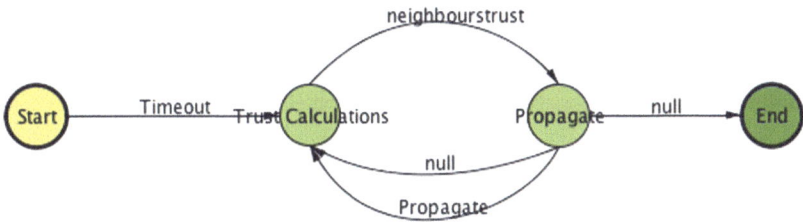

Fig. 3.7 The uHelp flooding protocol

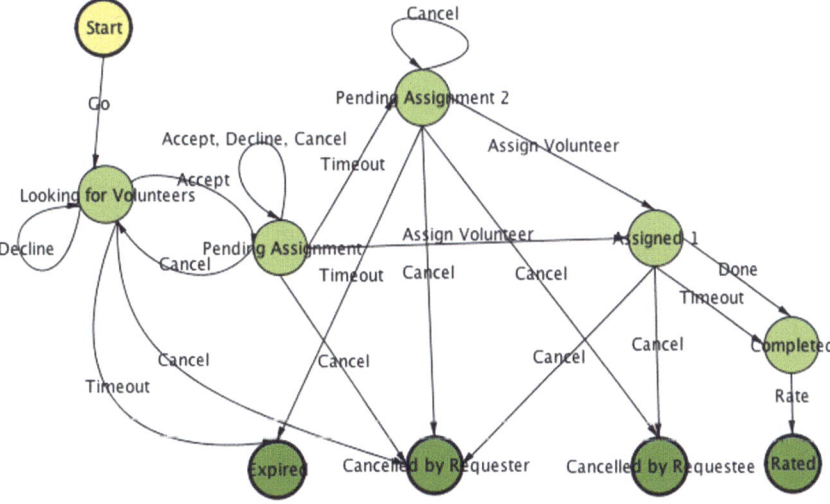

Fig. 3.8 The uHelp requests protocol

sending a help request, everyone will perform a stay and go to a new instance of the Flooding scene that deals with propagating this request. The protocol of the Help scene is straightforward and it is presented by Figure 3.6.

2. **Flooding scene**: In this scene, first the trust agent provides details about one's trust in one's neighbours, and the help request is then propagated (or not) accordingly. This repeats until propagation terminates, where the conditions for termination are discussed in further detail in Section 3.4.1. However, we note that when an instance of this scene is first created, the user agent issuing the help

request automatically performs a stay and go to a new instance of the Requests scene, to manage with request. Then, every time a new user agent receives this help request in the Flooding scene, and regardless of whether it propagates it or not, that user agent will also perform a stay and go to the corresponding instance of the Requests scene. The protocol of the Flooding scene is relatively straightforward and it is presented in Figure 3.7.

3. **Requests scene**: In this scene the Requestees can accept or decline, and they are allowed to change their mind, up until a certain deadline passes for collecting responses, which will move the scene to the next state (the "Pending Assignment 2" state of Figure 3.8). As long as there is at least one accept, and the deadline for assigning volunteers has not passed yet, the Requester will be able to select a volunteer (moving the state to "Assigned 1", Figure 3.8). At this point the Requestee either declares the task as done, or the task is assumed to be completed after the deadline passes (moving the state to "Completed", Figure 3.8). After this, the Requester is expected to rate the volunteer's performance, which triggers the trust agent to update the database accordingly. We note however, that both the Requester and Requestee may cancel the request, or volunteering, respectively, at any point in time (which moves the state to "Cancelled by Requester" or "Cancelled by Requestee", respectively – Figure 3.8). Additionally, if no Requestees volunteer in time, or the Requester does not select a volunteer in time, then the task will be marked as expired (the "Expired" state, Figure 3.8). The protocol of this scene is presented in Figure 3.8.

3.4 The Integrated Technologies

3.4.1 Flooding Algorithm

The flooding algorithm is the core computational process in the uHelp platform. It ensures requests for help are disseminated through the community. As stated earlier, we build upon a social network representation of the community: this is represented by some graph, in which the members of the community are nodes, and the edges represent friendship relations between two people. When someone wants to request help for a specific task, the flooding algorithm sends this request to that person's *trusted* neighbours in the graph (i.e. her friends) and from there it continues to flood through the network. This is similar to a number of other algorithms designed for rapidly disseminating a message through a graph, most prominently the Gnutella algorithm for P2P file sharing [12]. The main difference between existing approaches and the flooding algorithm discussed here is that the decision to stop forwarding the request is made primarily based on trust, rather than on other things, such as the time passed since the initial request was made. In fact, in Figure 3.1(a), one can see how the user controls the flooding algorithm by deciding the required trust level for a given task, along with the friendship level, which represents the number of hops in the graph (friends stand for one hop, friends of friends stand for two hops etc.).

The reason for this is that we not only want to find someone willing to volunteer for a task, but the person must also be trustworthy when performing this task. The way we calculate trust between two people is described in Section 3.4.3. To calculate trust along a path, we assume that trust satisfies the triangular norm inequality, that is for all α, β, γ nodes in a network $Trust(\alpha, \gamma) \leq T(Trust(\alpha, \beta), Trust(\beta, \gamma))$ for some T-norm function T. Thus, trust is monotonically decreasing along any path. We use multiplication for the experiments; for instance we get $Trust(\alpha, \gamma) = Trust(\alpha, \beta) \cdot Trust(\beta, \gamma)$, though more optimistic functions could be used, such as min. The flooding algorithm stops propagating the request when the cumulative trustworthiness of a node falls below a certain threshold τ. Algorithm 1 gives the pseudocode for the algorithm that performs this flooding.

Every node in the network runs Algorithm 1. It works in a straightforward manner: every time a node's user needs help she calls the FLOOD function with the following arguments:

- the $task$ she needs help with (which is obtained from the form the user fills in before pressing help),
- the minimum level of trust τ required in the person to execute the task (also obtained from the form the user fills in before pressing the "Help!" button; see the "Trust" field in Figure 3.1(a)),
- the maximum number of $hops$ (or friendship level) that the flooding algorithm is allowed (also obtained from the form the user fills in before pressing the "Help!" button; see the "Friendship" field in Figure 3.1(a)),
- the $deadline$ for someone to accept or reject (which is obtained from the task's deadline that the user specifies, see for instance the "Drop off" date of Figure 3.1(c). We note that this deadline precedes the task's deadline by a predefined fixed amount of time. For instance, if the deadline for the task is tonight at 6 p.m., then the deadline to accept might be today at 3 p.m., which is three hours before the task's deadline).

The other parameters of the FLOOD function are automatically set accordingly.

- $messagetype = HELP$, this is because the flooding algorithm is not only used to ask for help, but also to cancel previous requests. For instance, when the user cancels a request before selecting a volunteer, then all those who have been asked for help should be informed that the request is now cancelled. In this case, we run the same flooding algorithm, but with $messagetype = CANCELLED$. Also, when a user selects a volunteer, then all those that have been asked for help (except the selected volunteer) should be informed that their help is no longer needed. In this case, we again run the same flooding algorithm, but with $messagetype = NOTNEEDED$.
- $pathtrust = 1$, this is the initial trust. Every time a message propagates to its neighbouring nodes, trust is multiplied by the trust in that new node. We start with the value 1. That is, if the requester r propagates the request to its neighbour n, then r's trust in n is initially multiplied by 1.

Algorithm 1 Flooding Algorithm. We denote by $n \rightarrow m$ the request to execute method m at node n. Methods are defined as functions and are non-blocking

---

Require: $me : Node$ ▷ My identifier
Require: $friends : 2^{Node}$ ▷ Set of neighbouring nodes
Require: $Trust : Node \times Task \rightarrow [0, 1] \cup \bot$ ▷ Trust in friends for given tasks. $\bot$ for unknown people
Require: $OldPathTrust : Task \rightarrow [0, 1]$ ▷ Previous trust received from a path for given tasks. Initially it is -1.
Require: $T : [0, 1] \times [0, 1] \rightarrow [0, 1]$ ▷ A T-norm function, e.g. min, $\cdot$.
Require: $\sigma : [0, 1]$ ▷ Minimum increase in trust to re-flood the network
Require: $ReceivedRequests : 2^{Task}$ ▷ The set of received requests.

 1: **function** PROPAGATE($task, messagetype, \tau, pathtrust, path, deadline, hops$)
 2:   **if** $me \notin path$ and $Now() < deadline$ **then**
 3:     **if** $messagetype == HELP$ **then**
 4:       **if** $task \notin me.ReceivedRequests$ **then**
 5:         $OldPathTrust(task) := pathtrust$;
 6:         $me \rightarrow$ FLOOD($task, messagetype, \tau, pathtrust, path \oplus me, deadline, hops, false$);
 7:       **else if** $pathtrust - OldPathTrust(task) > \sigma$ **then**
 8:         $OldPathTrust(task) := pathtrust$;
 9:         $me \rightarrow$ FLOOD($task, messagetype, \tau, pathtrust, path \oplus me, deadline, hops, true$);
10:       **end if**
11:     **else**
12:       $me \rightarrow$ FLOOD($task, messagetype, \tau, pathtrust, path \oplus me, deadline, hops, false$);
13:     **end if**
14:   **end if**
15: **end function**
16: **function** FLOOD($task, messagetype, \tau, pathtrust, path, deadline, hops, asked?$)
17:   **if** $\neg asked?$ **then**
18:     **if** $messagetype == HELP$ **then**
19:       $me \rightarrow$ Msg_Help($task, deadline$);
20:     **else if** $messagetype == NOTNEEDED$ **then**
21:       $me \rightarrow$ Msg_NotNeeded($task$);
22:     **else if** $messagetype == CANCELLED$ **then**
23:       $me \rightarrow$ Msg_Cancelled($task$);
24:     **end if**
25:   **end if**
26:   **for all** $n \in friends$ **do**
27:     $NewPathTrust := T(Trust(n, task), pathtrust)$;
28:     **if** $NewPathTrust \geq \tau$ and $length(path) - 1 < hops$ **then**
29:       $n \rightarrow$ PROPAGATE($task, messagetype, \tau, NewPathTrust, path, deadline, hops$);
30:     **end if**
31:   **end for**
32: **end function**

- $path = \langle me \rangle$, this adds the requester as the first person in the path of people this
  message is being sent to. This ensures that the flooding algorithm does not later
  on ask the requester for help for his own request. (see Line 2 of Algorithm 1)
- $asked? = yes$, this is the parameter that decides whether the user should be
  informed of the new message or not, and it is needed to make sure that if one's
  uHelp app propagates a message more than once, the human user of this app is
  not informed more than once. We note here that we do allow a node to propagate
  a message more than once, if received from different paths. Re-flooding is
  discussed and motivated shortly.

When a node runs the FLOOD function, it first checks whether that user should be
informed of any new help requests (or that help is no longer needed, or cancelled, in
the other cases of using the flooding algorithm). Only if the user has not already been
informed (i.e. if $\neg asked?$), then the appropriate pop-up message is triggered. After
that, the node goes on to propagate that message to its neighbours, if the conditions
are fulfilled. Conditions are considered fulfilled when the trust in that neighbour is
above the threshold τ, and the number of hops within the acceptable limit ($hops$).
Recall that when adding a new node to the path, trust is updated by applying the
trust in the node propagating the request and the trust the propagating node has in
the propagated node (Line 27 of Algorithm 1).

The message is propagated by having the *trusted* neighbouring nodes executing the
PROPAGATE function. With the exception of the exclusion of the $asked?$ parameter,
the function has the same parameters as the FLOOD function, with the $pathtrust$
updated with the new trust value.

When a neighbouring node executes a PROPAGATE function, it first verifies that
there is no loop in the flooding by checking that the node is not already included
in the path, and that we are not beyond the deadline. Then, the node first checks
whether it has been asked for help. If it has not, then the FLOOD function is called,
with the $asked?$ parameter set to $false$ and the node appending itself to the $path$.
If it has been asked earlier for help for this specific task, then it is allowed to re-flood
the network if the task has been requested previously but the path trust in the node
is now sufficiently higher than before (by a difference larger than σ). In this case,
the FLOOD function is called, with the $asked?$ parameter set to $true$ and the node
appending itself to the $path$. This last point is very important as the depth to which
a request percolates the network depends on the cumulated level of trust. There may
be several paths connecting any two nodes, and thus the trust between them should
be taken to be the maximum of the cumulated trust over the paths. If we reach a node
with a certain level of trust we will flood the network with that value. However, if
later on a higher value is found we need to re-flood the network as this higher value
may make reachable nodes previously not reached due to the non-increasing effect
of the T-norm.

Concerning efficiency, we note that the flooding algorithm is not affected by
loops in the graph as flooding is stopped when the node itself is found in the
request path (due to the condition on Line 2 of Algorithm 1). Setting parameter
σ is very important for the efficiency of the algorithm, as if it is set very low the
number of messages can become exponential in the number of nodes for particular

topologies. In the worst case, with $\sigma = 0$, for a graph G, the number of messages is $\sum_{p \in loop\_free\_paths(G)} length(p)$, which can be exponential in the number of nodes. An adequate value has to be set experimentally.

Finally, we note that the algorithm limits the number of hops in addition to considering the trust level. This is achieved by comparing the length of the *path* with the *hops* parameter representing the maximum number of hops in the FLOOD method. This limitation on the number of hops makes our algorithm somehow similar to the Gnutella flooding algorithm, although in our case we also limit the flooding by taking trust into account.

3.4.2 Semantic Similarity

The tasks managed by the uHelp community consist of the *activity* the task is about (care for a relative, which could include babysitting, picking up, cooking etc.), and sometimes, the relevant *relative* to whom the activity is applied (we focus on children, and hence we may have baby, toddler, preschooler etc.). Consequently we need to handle two separate hierarchies.

On one hand, the different kinds of activities are usually organised in a meronomy,[3] specifying which sub activities may be part of a given activity (e.g. that changing nappies is a sub activity of babysitting). Here we take inspiration from the design decisions suggested by the Process Specification Language (PSL) [6]. An example of a meronomy of activities and sub activities for the uHelp community is given in Figure 3.9.

Fig. 3.9 Meronomy of activities

On the other hand, the different kinds of children are more naturally organised in a taxonomy, specifying subclasses of children (e.g. that a toddler is a child). Figure 3.10 shows an example of a taxonomy of children for the uHelp community that reflects a hypernym-hyponym relation, which is similar to the one defined in WordNet [4].

---

[3] A meronomy is a type of hierarchy that deals with part-of relationships, in contrast to a taxonomy whose structure is based on is-a relationships.

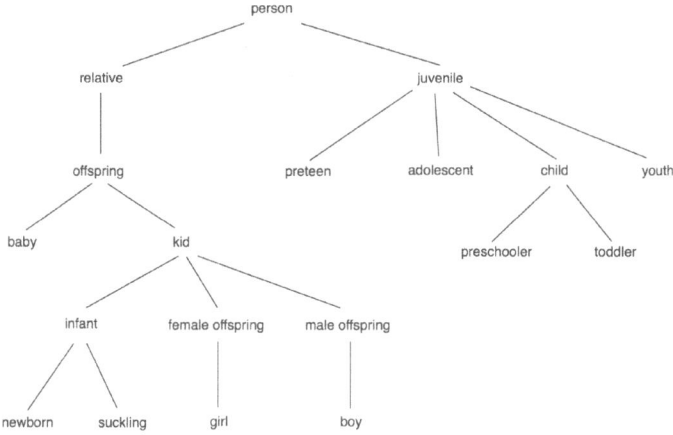

Fig. 3.10 Taxonomy of children

For our specific case study, a task is a pairing of a "care for" activity with a "child", for instance ⟨*giving a lift, preschooler*⟩ or ⟨*playing, toddler*⟩. If the activity specified is not a leaf of the meronomy, the requester expects the volunteer to be capable of performing all the subactivities involved. For instance, if the activity is *giving a lift*, the volunteer is expected to perform both *picking up* and *dropping off*. However, in a taxonomy this is treated differently: if the child is not a leaf of the taxonomy it is simply not as detailed a specification: a schoolchild is either a preteen or an adolescent, but clearly not both.

This gives the requester the freedom to specify a task at a more, or less, detailed level, which can affect how the flooding algorithm propagates the request. Specifically, the specification of the task has a direct effect on the trust model, because any evaluation of a volunteer's performance is linked to the task he performed. When evaluating a target for the performance of another task, the *similarity* between the new task and previously performed tasks is an important factor to take into account for estimating his performance.

Consequently, task similarity has to reflect this influence on the quality of task performance. For this reason, activities and their meronomical structure, and types of children and their taxonomical structure, affect the trust evaluations of specific tasks in different ways.

We use OpinioNet [11] to propagate evaluations of other activities through the meronomy. In OpinioNet, opinions may be assigned by users to nodes of a structural graph, and the OpinioNet algorithm gives a method for propagating these opinions throughout the graph. In uHelp, the opinions are satisfaction ratings of how a user performed an activity and the structural graph is the meronomy of those activities. We argue that a user's performance of one activity should influence his trustworthiness for performing similar activities (nearby nodes in the meronomy). We make use of the OpinioNet algorithm for the propagation of trust in graphs based on the part-of relation.

The basic idea of the OpinioNet algorithm is that if a node in the graph does not receive a direct evaluation, then its evaluation may be deduced from its children nodes' evaluations. This is because the parent node is structurally composed of its children nodes. Hence, the evaluations of children nodes must necessarily influence the deduced evaluation of a parent node. OpinioNet refers to the direct evaluation of a node or an evaluation of it that is deduced from evaluations of the parts that compose it as the 'intrinsic evaluation' of that node.

Additionally, an 'extrinsic evaluation' is an evaluation that is propagated down from parent nodes to children. In the absence of information about the node itself, or the parts that compose it, information may be inherited from what one belongs to. In other words, in the absence of information about intrinsic evaluations, the evaluation of that node is calculated based on evaluations of its parents' nodes.

As an example of how the OpinioNet algorithm propagates evaluations through the meronomy, consider that if the target performed well at the *preparing meal* activity, this will affect the evaluation of that target for the activity of *babysitting*. This is an 'intrinsic evaluation'. An 'extrinsic evaluation' would be to say that performing well at *babysitting* carries over to a good evaluation for the subactivity of *changing nappies* if there is no direct evaluation of the target performing the *changing nappies* activity.

However, we also argue that many of the activities change, dependent on the type of child. For instance, preparing a meal for a baby is very different to preparing a meal for an adolescent. We therefore need to take the similarity between children into account as well.

The similarity over the child taxonomy is considered from a more semantic perspective and we use the measure proposed by Li et al. [10], which combines well-known edge-based and node-based techniques, and correlates well with similarity assessments as done by humans. In particular, it takes three aspects of taxonomies into account:

- the distance l between two tasks t_1 and t_2 in the taxonomy: the closer, the more similar the tasks are;
- the depth h in the taxonomy of their most specific subsumer $sub(t_1, t_2)$: the deeper in the taxonomy the subsumer is, the more similar the tasks are;
- the local semantic density d of instances of these tasks: the greater the information content of the subsumer, $d = -\log p(sub(t_1, t_2))$, the more similar the tasks are.

Consequently, Li et al. define the first measure to be anti-monotonic in the range [0, 1] and the other measures to be monotonic in the same range, where 0 represents complete dissimilarity, and 1 represents complete similarity. The contribution of each of these measures is taken independently of each other:

$$sim(t_1, t_2) = e^{-\alpha l} \cdot \tanh \beta h \cdot \tanh \lambda d$$

Constants α, β and λ are positive real numbers that determine the relative influence of each of the three measures on the final similarity. They provide us with suitable adjustment points for the overall semantic similarity to fit well and evolve with the actual usage of terms by a concrete community.

3.4.3 Trust Model

Every member of the community maintains his own trust evaluations of other members of the community. These trust evaluations are task-dependent, by which we mean that a user's evaluation of another may change, depending on the task for which he is evaluated. This trust evaluation is crucial in the functioning of the flooding algorithm, which only forwards a request for help to a neighbour if the trust level in that neighbour is sufficiently high. The transitivity of trust is not uncommon in the trust literature. For instance, TidalTrust [5] is a trust model that propagates trust evaluations through a social network, although a more sophisticated algorithm is used to compute the trustworthiness of a node. For a more in-depth discussion on transitivity of trust and considerations that should be taken into account when propagating trust through a network, we refer the interested reader to [7, 3].

The trust mechanism implemented in uHelp makes use of the evaluations that users generate after a task is completed by a volunteer.

For instance, every time someone volunteers to care for a child, the adult responsible for the child generates an evaluation. An evaluation E is a tuple with the form

$$(Requester, Volunteer, ActivityType, ChildType, Value)$$

where $Requester$ is the person that was asking for help, $Volunteer$ is the ID of the person that took care of the child, $ActivityType \in$ Meronomy of activities, $ChildType \in$ leafs in Children's taxonomy, and $Value \in [0, 1]$ with 0 meaning complete failure and 1 complete satisfaction. We refer to the elements in the tuple using superscripts. For instance, to refer to the ChildType value we use the form $E^{ChildType}$.[4]

If the activity includes other subactivities, the user (the requester of the task) can decide to evaluate the general activity only, or to perform a detailed evaluation of each one of the leaf activities. For instance, if the activity was "giving a lift", the user can evaluate just "'giving a lift" or he/she can evaluate each one of the single subactivities ("picking up", "dropping off"). The ChildType is always a leaf in the Children's taxonomy.

Some examples of evaluations:

- $(Ann123, John243, "entertaining", "toddler", 0.7)$
- $(Paul324, Mary334, "singing songs", "preschooler", 0.8)$

---

[4] Note that although in this section we talk about caring for children, the same approach is followed for other tasks, where $ActivityType$ depends on the task, and the $ChildType$ can be replaced either by $ObjectType$ (for instance when someone asks others to donate to them, lend them or get them something) or $LocationType$ (for instance when someone asks others to go with them somewhere).

We calculate the trust in an individual as a function of the trust that the individual has earned regarding the type of activity (see meronomy of activities) and the object of the activity (see Children's taxonomy).

$$Trust(R, V, T, Ch, t) = \alpha * Trust_{object}(R, V, Ch, t) + (1-\alpha) * Trust_{activity}(R, V, T, t)$$

where R is the requester ID, V the volunteer ID, T the activity, Ch the type of child, t the time and α the weight given to each type of trust. We discuss each type of trust next.

$Trust_{object}(R, V, Ch, t)$. We use a similarity threshold to filter out evaluations associated with types of children too distant from the target child in the Children's taxonomy. The similarity value for the Children's taxonomy is calculated using the following formula:

$$simVal(R, Ch_i, Ch_j) = \begin{cases} sim(Ch_i, Ch_j) & if \, sim(Ch_i, Ch_j) > simTh \\ 0 & otherwise \end{cases}$$

where $sim(Ch_i, Ch_j) \in [0, 1]$ is the semantic similarity between types of children Ch_i and Ch_j as already defined in the previous section and $simTh \in [0, 1]$ is the similarity threshold defined by the community. Below that threshold, the similarity between two types of children is considered too low for the associated experience to be useful.

The trust that requester R places in volunteer V associated with a particular object Ch at time t is calculated using the formula:

$$Trust_{object}(R, V, Ch, t) = \frac{\sum_{E_i \in Evaluations(t)} (simVal(R, Ch, E_i^{ChildType}) \cdot E_i^{Value})}{\sum_{E_i \in Evaluations(t)} simVal(R, Ch, E_i^{ChildType})}$$

where $Evaluations(t)$ is the subset of evaluations in the time window $[t - TWin, t]$, where $TWin$ is a predefined parameter.

$Trust_{activity}(R, V, T, t)$. To evaluate the trust in an individual regarding the type of activity that she/he is requested to perform, we make use of the meronomy of activities and the OpinioNet algorithm [11]. OpinioNet allows us to consider not only the direct evaluations of an activity but the evaluations associated with subactivities and superactivities. For instance, if there are evaluations of a volunteer associated with "entertaining" a child, these evaluations can be used to say something about the capacity of the volunteer regarding "singing songs", "playing" and "watching movies". Similarly, if there is an evaluation associated with "dropping off" a child it can be used to say something about the capacity of the volunteer at "accompanying" a child. For the details of how OpinioNet works, we refer the interested reader to [11].

References

1. Beech, S., Geelhoed, E., Murphy, R., Parker, J., Sellen, A., Shaw, K.: The lifestyles of working parents: Implications and opportunities for new technologies. Tech. Rep. HPL-2003-88(R.1), HP Laboratories (2004)

2. d'Inverno, M., Luck, M., Noriega, P., Rodríguez-Aguilar, J.A., Sierra, C.: Communicating open systems. Artificial Intelligence **186**(0), 38–94 (2012). DOI 10.1016/j.artint. 2012.03.004. URL http://www.sciencedirect.com/science/article/pii/ S0004370212000252

3. Falcone, R., Castelfranchi, C.: Trust and transitivity: a complex deceptive relationship. In: Proceedings of the 15th International Workshop on Trust in Agent Societies, co-located with AAMAS 2010, pp. 43–53 (2010)

4. Fellbaum, C.: WordNet: An Electronic Lexical Database. Language, Speech, and Communication. MIT Press, Cambridge (1998). URL https://books.google.es/books?id= Rehu8OOzMIMC

5. Golbeck, J.: Combining provenance with trust in social networks for semantic web content filtering. In: L. Moreau, I. Foster (eds.) Proceedings of the 2006 International Conference on Provenance and Annotation of Data, IPAW'06, *Lecture Notes in Computer Science*, vol. 4145, pp. 101–108. Springer, Berlin (2006). DOI 10.1007/11890850_12. URL http://dx.doi. org/10.1007/11890850_12

6. Grüninger, M., Menzel, C.: The process specification language (PSL) theory and applications. AI Mag. **24**(3), 63–74 (2003). URL http://dl.acm.org/citation.cfm?id= 958671.958677

7. Jøsang, A., Gray, E., Kinateder, M.: Simplification and analysis of transitive trust networks. Web Intelli. and Agent Sys. **4**(2), 139–161 (2006). URL http://dl.acm.org/citation. cfm?id=1239776.1239778

8. Koster, A., Madrenas-Ciurana, J., Osman, N., Schorlemmer, W.M., Sabater-Mir, J., Sierra, C., Fabregues, A., de Jonge, D., Puyol-Gruart, J., Garcia-Calvés, P.: u-Help: supporting helpful communities with information technology. In: M.L. Gini, O. Shehory, T. Ito, C.M. Jonker (eds.) Proceedings of the 12th International Conference on Autonomous Agents and Multi-Agent Systems (AAMAS 2013), pp. 1109–1110. IFAAMAS, Richland (2013). URL http: //dl.acm.org/citation.cfm?id=2485095

9. Koster, A., Madrenas-Ciurana, J., Osman, N., Schorlemmer, W.M., Sabater-Mir, J., Sierra, C., de Jonge, D., Fabregues, A., Puyol-Gruart, J., Garcia-Calvés, P.: u-Help: Supporting helpful communities with information technology. In: S. Ossowski, F. Toni, G.A. Vouros (eds.) Proceedings of the First International Conference on Agreement Technologies (AT 2012), *CEUR Workshop Proceedings*, vol. 918, pp. 378–392. CEUR-WS.org (2012). URL http: //ceur-ws.org/Vol-918/111110378.pdf

10. Li, Y., Bandar, Z.A., McLean, D.: An approach for measuring semantic similarity between words using multiple information sources. IEEE Transactions on Knowledge and Data Engineering **15**(4), 871–881 (2003)

11. Osman, N., Sierra, C., Sabater-Mir, J.: Propagation of opinions in structural graphs. In: H. Coelho, R. Studer, M. Wooldridge (eds.) Proceedings of the 19th European Conference on Artificial Intelligence (ECAI 2010), *Frontiers in Artificial Intelligence and Applications*, vol. 215, pp. 595–600. IOS Press, Amsterdam (2010)

12. Ripeanu, M.: Peer-to-peer architecture case study: Gnutella network. In: Proceedings First International Conference on Peer-to-Peer Computing, pp. 99–100 (2001). DOI 10.1109/P2P. 2001.990433

13. Sellen, A., Hyams, J., Eardley, R.: The everyday problems of working parents: Implications for new technologies. Tech. Rep. HPL-2004-37, HP Laboratories (2004)

Chapter 4
The WeCurate Application

Matthew Yee-King, Dave de Jonge, Roberto Confalonieri, Nardine Osman, Mark d'Inverno, Carles Sierra, Leila Amgoud and Katina Hazelden

Multiuser museum interactives are computer systems installed in museums or galleries that allow several visitors to interact together with digital representations of artefacts and information from the museum's collection. WeCurate is such a system that allows users to collaboratively create a virtual exhibition from a cultural image archive. It provides a synchronised image browser across multiple devices to enable a group of users to work together to curate a collection of images. WeCurate uses electronic institutions to coordinate and synchronise the interactions between individuals, and it relies on agreement technologies (such as argumentation and computational social choice) for collective decision making. This paper provides an overview of the WeCurate application, describes its underlying electronic institution, and presents a brief introduction to its collective-decision-making mechanism.

Matthew Yee-King
Goldsmiths College, London, e-mail: m.yee-king@gold.ac.uk

Dave de Jonge
IIIA-CSIC, Barcelona, e-mail: davedejonge@iiia.csic.es

Roberto Confalonieri
University of Padua, Padua, e-mail: roberto.confalonieri@unipd.it

Nardine Osman
IIIA-CSIC, Barcelona, e-mail: nardine@iiia.csic.es

Mark d'Inverno
Goldsmiths College, London, e-mail: dinverno@gold.ac.uk

Carles Sierra
IIIA-CSIC, Barcelona, e-mail: sierra@iiia.csic.es

Leila Amgoud
IRIT-CNRS, Toulouse, e-mail: Leila.Amgoud@irit.fr

Katina Hazelden
Crown Commercial Service, London, e-mail: kat9@mac.com

© Springer Nature Switzerland AG 2024
N. Osman (ed.), *Electronic Institutions*, Artificial Intelligence: Foundations, Theory, and Algorithms, https://doi.org/10.1007/978-3-319-65605-2_4

4.1 Objectives

In recent times, high-tech museum interactives have become ubiquitous in major institutions. Typical examples include augmented-reality systems, multi-touch table tops and virtual reality tours [11, 14, 20]. Whilst multiuser systems have begun to appear, e.g. a 10-user quiz game in the Tate Modern, the majority of these museum interactives do not perhaps facilitate the sociocultural experience of visiting a museum with friends, often being designed for a single user. At this point, we should note that mediating and reporting the actions of several 'agents' to provide a meaningful and satisfying sociocultural experience for all is challenging, requiring multiple-criteria decision making. Another trend in museum curation is the idea of community curation, where a community discourse is built up around the artefacts, to provide different perspectives and insights [19]. This trend is not typically represented in the design of museum interactives, where information *browsing*, not information *generation*, is the focus. However, museums are engaging with the idea of crowdsourcing with projects such as 'Your Paintings Tagger' and 'The Art Of Video Games' [12, 6]. Again, controlling the workflow within a group to engender discussion and engagement with the artefacts is challenging, especially when the users are casual ones as in a museum context.

In this chapter, we describe WeCurate [22, 2], an image browser for collaboratively curating a virtual exhibition from a cultural image archive – group curation. The WeCurate application allows users to synchronously view media and enables negotiation about which images should be added to the group's image collection. Further, it accelerates the navigation of extensive museum databases and provides a platform for sociocultural experiences, combining the actions of autonomous agents and users to facilitate decision making. WeCurate is tasked with establishing the users' presence in the shared experience by enabling communication around the deconstruction and appropriateness of the media, representing social proxies, and making agents' and group members' actions manifest to everyone.

WeCurate uses a multiagent system to support community interactions and decision making and an electronic institution [9] to model the workflow. The aim of this work is not only to make use of agent technology and electronic institutions as a means to implement a multiuser museum interactive, but also to relate agent theory to practice in order to create a usable system that will help us to answer the following research questions. Do agent technology and electronic institutions enable users to share online experiences in a way that was not possible before? Are decision-making capabilities and aggregation operators defined in the literature suitable for enhancing the experience of users? Can we create a user interface that represents the state of the underlying system to the user in a meaningful way? Can we create a sense of social presence between the users of the system?

To this end, we specify a community curation session in terms of the scenes of an electronic institution for controlling community interactions. We make use of a multimodal user interface which directly represents users as agents in the scenes of an underlying electronic institution and which is designed to engage casual users in a social discourse around museum artefacts. One key underlying technology is then used to support the system's and users' decisions by means of personal assistant agents equipped with different decision-making capabilities.

In WeCurate, the automated decision making of the agents is undertaken by different models [8]: *preference aggregation* [23], *multiple-criteria decision making* [1, 17] and *voting*. Preference aggregation allows a quick understanding of whether the users consider an image interesting or not. To this end, each user expresses an *image preference* and a collective decision is achieved by preference aggregation. Multiple-criteria decision making is a collective decision taken as a result of a negotiation protocol. Users exchange *image arguments* according to an *argument-based multiple-criteria decision-making* protocol. Voting takes the majority of the agents' decisions into account. For further information on automated decision-making technology, we refer the interested reader to [8].

4.2 Illustration

In this section we present the user interface of the WeCurate system [13]. The interface is an animated user interface with a separate screen for each scene in the electronic institution. The three main scenes are presented by the three main views in Figures 4.1, 4.2 and 4.3.

1. **Selection view.** The purpose of this view is to allow a quick decision as to whether an image is interesting enough for a full discussion. Users can *zoom* into the image and see the zooming actions of other users. They can also set their *overall preference* for the image using a simple like/dislike slider, which also shows the preferences of other users in the group. The view is shown in Figure 4.1.
2. **Forum view.** If an image is deemed interesting enough by the agents, the users are taken to the forum view, where they can engage in a full *discussion* of the image. Users can add, delete and weight tags, and can see the actions of the other users so they have a sense of what others in the group are interested in. They can also view images that were previously added to the collection. The view is shown in Figure 4.2.
3. **Voting view.** In this view, the users can vote on whether they wish to store the image to their collection. Here, the decision is made to add an image to the group collection, or not. The view is shown in Figure 4.3.

4.3 The EI Specification

In this section we show the underlying electronic institution (EI) specification [9] of the WeCurate application.

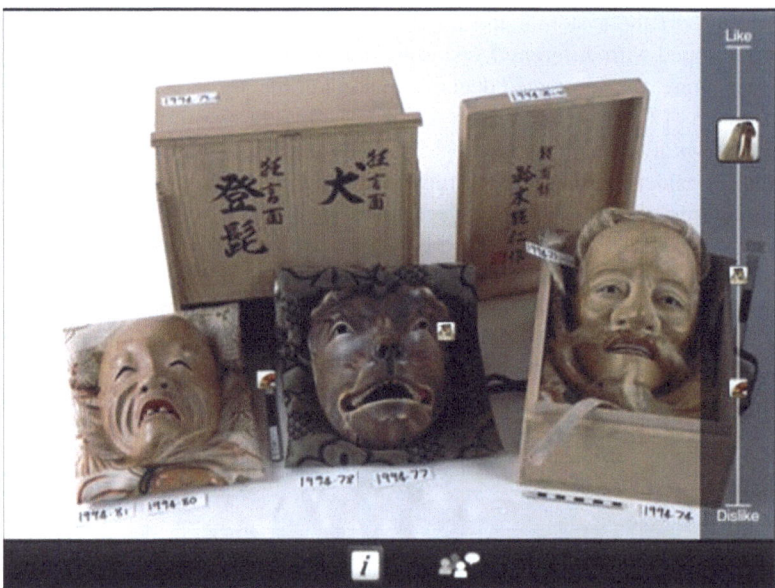

Fig. 4.1 "Selection" view, which aims to gauge the interest level of users in the proposed image

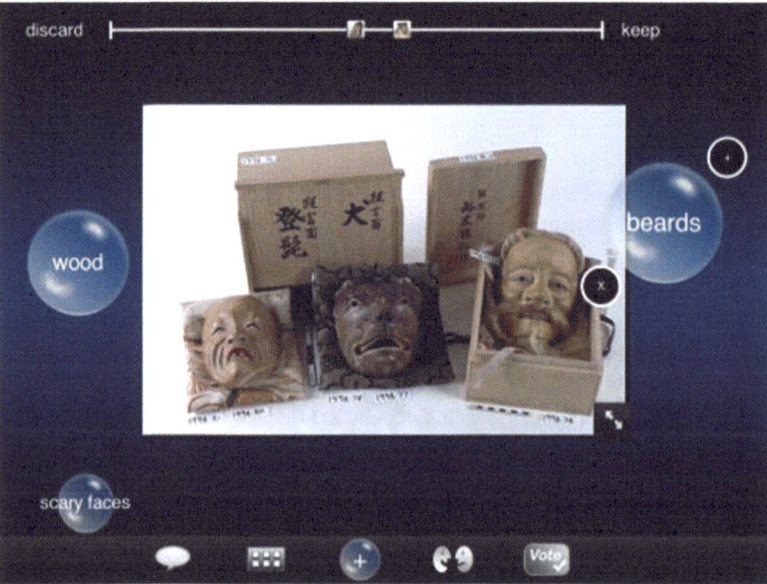

Fig. 4.2 "Forum" view, which allows users to engage in a discussion about the image, once it has been deemed interesting in the selection scene

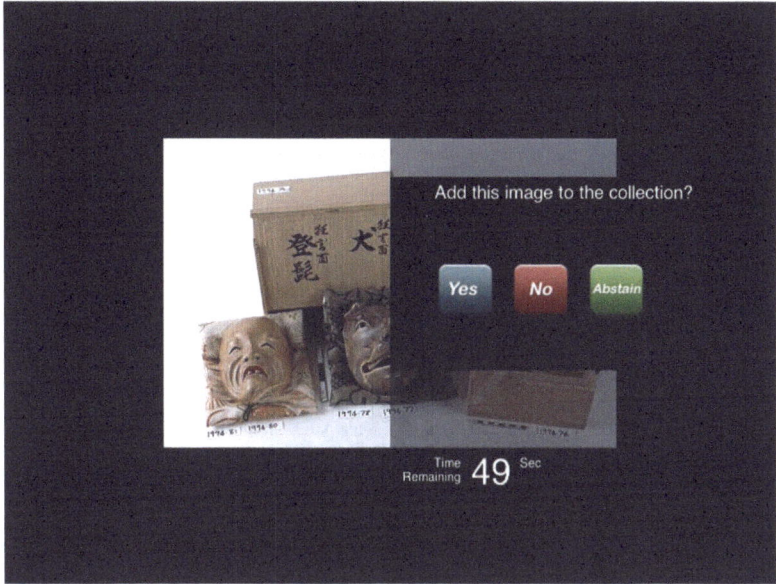

Fig. 4.3 "Voting" view, where users can vote on whether they wish to store the image to their collection

4.3.1 Dialogical Framework

In the EI model of the WeCurate application, we can distinguish three roles:

1. **User Assistant**: the agent that represents the user in the institution.
2. **Media Provider**: the agent that sends the image files to the users.
3. **Manager**: the agent that is responsible for creating the scene instances.

The institution should contain exactly one agent playing the role of Media Provider and exactly one agent playing the role of Manager. Furthermore there is one agent playing the role of User Assistant for each user participating in the application.

4.3.2 Performative Structure

The workflow of WeCurate can be divided into three stages: selecting interesting images, discussing selected images and voting on disputed images. These three stages can naturally be represented in an EI specification as three separate scenes. Each of these scenes will have exactly one instance. All User Assistants should always be together in the same scene instance. Note that, as illustrated by Figure 4.4, the Media Provider only enters the Selection scene, and this agent stays in that scene

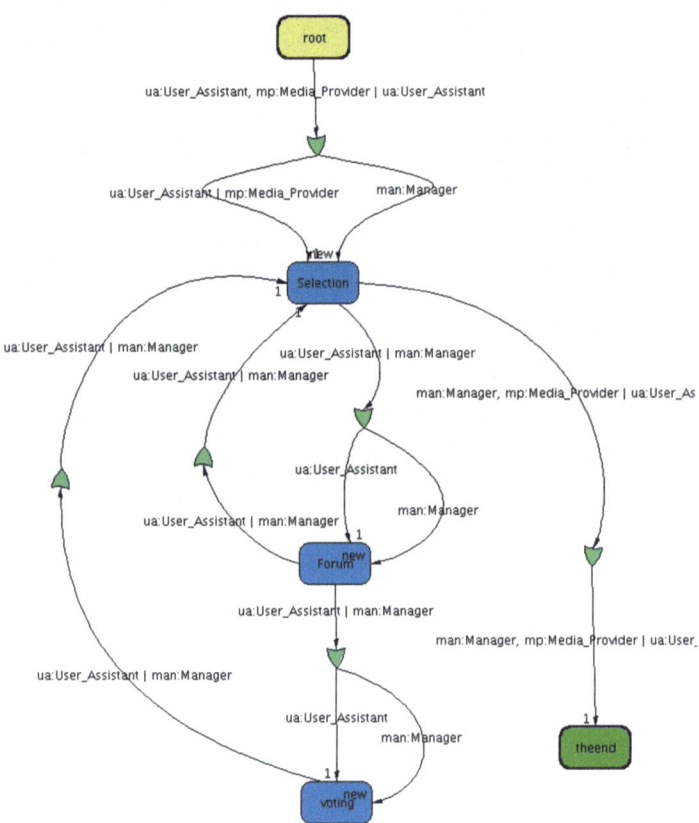

Fig. 4.4 The performative structure of the WeCurate institution

until the institution closes. The performative structure of the institution is presented
in Figure 4.4. Next, we present the three scenes in more detail.

1. **Selection**: The Selection scene, whose protocol is illustrated by Figure 4.5, is the
   first scene the users enter when they enter the WeCurate institution. The protocol
   starts with the server sending an image to the users. In an Electronic Institution
   we can represent this by the agent playing the role of Media Provider sending a
   message to the User agents containing a URL of the image on the server. Note
   that in the EI infrastructure it is not possible to send files with messages, so
   we cannot send the image itself. In the application the users can then express
   their opinion about the image by moving a slider up or down, where the slider

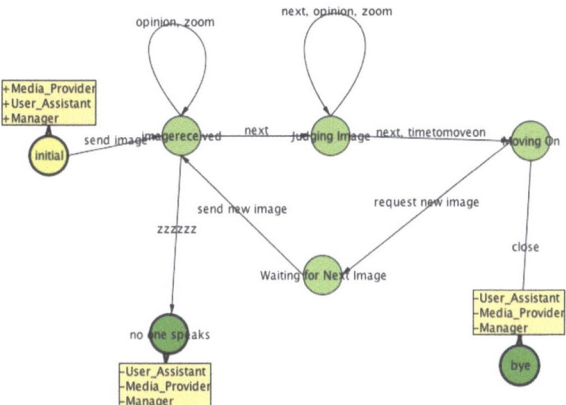

Fig. 4.5 The WeCurate selection protocol

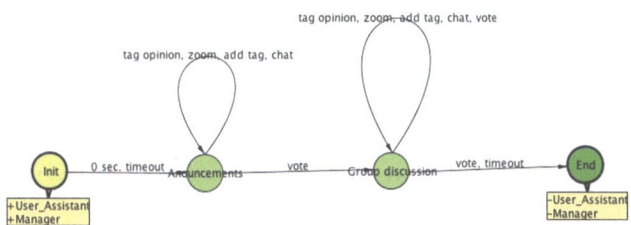

Fig. 4.6 The WeCurate forum protocol

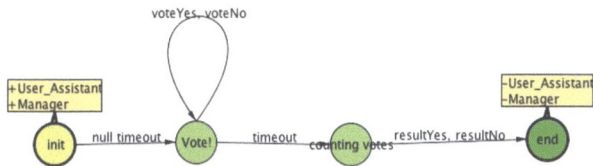

Fig. 4.7 The WeCurate voting protocol

being down means that the user totally dislikes the image, and the slider being up means he or she totally likes the image. In the Electronic Institution, this is modelled as a User Assistant sending a message to all the other User Assistants containing a number between 0 and 1, which represents the user's preference level.

Also users can zoom in on parts of the image. Whenever a user zooms in, the other users are able to see on their screens where the first user is zooming. In the institution, this is modelled by means of a 'zoom' message sent to all other User Assistants that contains the coordinates of the location where the user is zooming.

Once a user has made his or her decision about the image, he or she can click 'next'. When the first user clicks 'next' a timer starts counting down from 10 to 0. The scene then continues until either all other users have also clicked 'next' or the timer expires. Once you have clicked 'next' you cannot do anything anymore in the scene so you have to wait until the protocol finishes. We have modelled this by introducing two states in which the users can zoom. Whenever the first user sends a 'next' message the protocol moves from the first to the second state. The difference between these two states is that the second state contains an arc with a timeout, to represent the countdown.

Next, one of two things may happen: either the Manager decides that the image was not interesting enough for further consideration, in which case a new image will be requested from the Media Provider, or the users move on to the Forum scene for further discussion.

2. **Forum**: In the Forum scene, whose protocol is illustrated in Figure 4.6, the users further 'discuss' the image from the previous scene. They can not only express their overall impression of the image, but they can also express which aspects of the image they like or do not like. In the real application they do this by adding tags to the image; by increasing or decreasing the size of the tags they give more or less weight to those tags. For example if you like an image especially because it has a lot of colours, then you add the tag 'colours' and make it very large. In the Electronic Institution, we represent this as a message containing a string and a number, representing the tag itself and its size respectively. This continues until all the users have indicated they they want to move on to the Voting scene, or until a timer expires.

3. **Voting**: Finally, when the users are in the voting scene they can vote whether they want the image to be added to their shared photo album or not. In the Electronic Institution, this is represented by a message containing one of the three values: 'yes', 'no', or 'abstain'. After voting the users move back to the Selection scene. The protocol of the Voting scene is presented in Figure 4.7.

4.4 Integrated Technologies: Automated Decision Making

As previously mentioned in Sect. 4.1, the automated decision making of the agents is undertaken by different models: *preference aggregation* [23], *multiple-criteria decision making* [1, 17] and *voting*. In the following sections we describe these decision models and how they have been adopted in the different scenes of the WeCurate system.

4.4.1 Selection Scene and Preference Aggregation

The main goal of each user running in a selection scene is to express a preference about the image currently browsed. When the scene is finished, the User Assistant agents compute an evaluation of the image — the *image's interestingness* to the group of users — based on preference aggregation in order to decide whether to proceed with a Forum scene or to go back to a Selection scene (with a different image).

Let $\mathcal{I} = \{im_1, \ldots, im_n\}$ be the set of available images where each $im_j \in \mathcal{I}$ is the identifier of an image. The *image preference* of a user w.r.t. an image is a value that belongs to a finite bipolar scale $\mathcal{S} = \{-1, -0.9, \ldots, 0.9, 1\}$ where -1 and $+1$ stand for "reject" and "accept" respectively. Given a group of users $\mathcal{U} = \{u_1, u_2, \ldots, u_n\}$, we denote the image preference of a user u_i w.r.t. an image im_j by $r_i(im_j) = v_i$ with $v_i \in \mathcal{S}$.

A preference aggregator operator for merging the preferences of a group of n users $\{u_1, u_2, \ldots, u_n\}$ w.r.t an image im_j is defined as a mapping $f_{agg} : \mathcal{S}^n \to \mathcal{S}$. Then, a decision criterion for the collective decision about the interestingness of an image im_j can be defined as

$$\text{int}(im_j) = \begin{cases} 1, & \text{if } 0 < f_{agg}(\mathbf{r}) \leq 1 \\ 0, & \text{if } -1 \leq f_{agg}(\mathbf{r}) \leq 0 \end{cases} \tag{4.1}$$

where $\mathbf{r} = \{r_1(im_j), \ldots, r_n(im_j)\}$ is a vector consisting of the image preferences of n users w.r.t. an image im_j.

In the Selection scene, f_{agg} and $\text{int}(im_j)$ are instantiated by three different preference aggregators, as we will see.

4.4.1.1 Image Interestingness Based on Arithmetic Mean

A straightforward way to aggregate image preferences is by arithmetic mean. The *image interestingness* of an image im_j to a group of users $\{u_1, u_2, \ldots, u_n\}$ based on the arithmetic mean, denoted by $\overline{f(\mathbf{r})}$, and its respective decision criterion are defined as

$$\overline{f(\mathbf{r})} = \frac{\sum_{1 \leq i \leq n} r_i}{n} \tag{4.2}$$

$$\overline{\mathtt{int}}(im_j) = \begin{cases} 1, & \text{if } 0 < \overline{f(\mathbf{r})} \leq 1 \\ 0, & \text{if } -1 \leq \overline{f(\mathbf{r})} \leq 0 \end{cases} \tag{4.3}$$

Therefore, the system proceeds with a Forum scene when $\overline{\mathtt{int}}(im_j) = 1$, while the system goes back to a Selection scene when $\overline{\mathtt{int}}(im_j) = 0$.

Since the arithmetic mean cannot take users' activity into account, we define two preference aggregators which can consider the users' engagement in the select scene.

4.4.1.2 Image Interestingness Based on Weighted Mean

In this case, each User Assistant agent also stores the zoom activity of its user. The zoom activity is a measure of the user's interest in a given image and, as such, it can be used to calculate the importance or weight of the image preference of each user. To this end, let us denote the number of *image zooms* of user u_i w.r.t. an image im_j as $z_i(im_j)$. Then, we can define the total number of zooms for an image im_j as $z(im_j) = \sum_{1 \leq i \leq n} z_i(im_j)$. Based on $z(im_j)$ and the z_i's associated with each user, we can define a weight for the image preference r_i of user u_i as $w_i = z_i(im_j)/z(im_j)$.

Therefore, the *image interestingness* of n users w.r.t. an image im_j based on weighted mean, denoted by $\overline{f_{wm}(\mathbf{r})}$, and the respective decision criterion can be defined as

$$\overline{f_{wm}(\mathbf{r})} = \frac{\sum_{1 \leq i \leq n} r_i w_i}{\sum_{1 \leq i \leq n} w_i} \tag{4.4}$$

$$\overline{\mathtt{int}_{wm}}(im_j) = \begin{cases} 1, & \text{if } 0 < \overline{f_{wm}(\mathbf{r})} \leq 1 \\ 0, & \text{if } -1 \leq \overline{f_{wm}(\mathbf{r})} \leq 0 \end{cases} \tag{4.5}$$

4.4.1.3 Image Interestingness Based on WOWA Operator

An alternative criterion for deciding whether an image is interesting or not can be defined by using a richer average operator such the Weighted Ordered Weighted Average (WOWA) operator [18]. The WOWA operator is an aggregation operator which allows us to combine some values according to two types of weights: i) a weight referring to the importance of a value itself (as in the weighted mean), and ii) an ordering weight referring to the values' order. Indeed, WOWA generalizes both the weighted average and the ordered weighted average [21]. Formally, WOWA is defined as [18]

$$f_{wowa}(r_1, \ldots, r_n) = \sum_{1 \leq i \leq n} \omega_i r_{\sigma(i)} \tag{4.6}$$

where $\sigma(i)$ is a permutation of $\{1, \ldots, n\}$ such that $r_{\sigma(i-1)} \geq r_{\sigma(i)} \; \forall i = 2, \ldots, n$, ω_i is calculated by means of an increasing monotone function $w^*(\sum_{j \leq i} p_{\sigma(j)}) - w^*(\sum_{j < i} p_{\sigma(j)})$, and $p_i, w_i \in [0, 1]$ are the weights and the ordering weights associated with the values respectively (with the constraints $\sum_{1 \leq i \leq n} p_i = 1$ and $\sum_{1 \leq i \leq n} w_i = 1$).

We use the WOWA operator for deciding whether an image is interesting in the following way. Let us take the weight p_i for the image preference r_i of user u_i as the percentage of zooms made by the user (like above). As far as the ordering weights are concerned, we can decide to give more importance to image preference values closer to extreme values such as -1 and $+1$, since it is likely that such values can trigger more discussions among the users rather than image preference values that are close to 0. Let us denote the sum of the values in $S^+ = [0, +0.1, \ldots, +0.9, +1]$ as s. Then, for each image preference $r_i(im_j) = v_i$ we can define an ordering weight as $w_i = r_i(im_j)/s$. Please notice how the p_i's and w_i's defined satisfy the constraints $\sum_{1 \leq i \leq n} p_i = 1$ and $\sum_{1 \leq i \leq n} w_i = 1$.

Then, a decision criterion for the interestingness of an image im_j based on $f_{wowa}(r_1, \ldots, r_n)$ can be defined as

$$\texttt{int}_{\text{wowa}}(im_j) = \begin{cases} 1, & \text{if } 0 < f_{wowa}(\mathbf{r}) \leq 1 \\ 0, & \text{if } -1 \leq f_{wowa}(\mathbf{r}) \leq 0 \end{cases} \tag{4.7}$$

4.4.2 Forum Scene and Argument-Based Multiple-Criteria Decision Making

The main goal of the users in a Forum scene is to discuss an image by pointing out what they like or dislike about the image through *image arguments* based on tags. During the tagging, each user is associated with an *overall image preference* that is automatically updated. Whilst tagging is the main activity of this scene, users can also engage in bilateral discussions with the aim of reaching mutual agreements about keeping or discarding an image. When a user is tired of tagging, he can propose to the other users to move to a Voting scene. In this case, an automatic *multi-criteria decision* is taken in order to decide whether the current image can be added or not to the image collection without a vote being necessary.

In our system each image is described with a finite set of *tags* or *features*. In what follows, we show how weighted tags, that is, tags associated with a value belonging to a bipolar scale, can be used to define arguments for or against a given image and to specify a multiple-criteria decision-making protocol to let a group of users decide whether or not to accept the image.

4.4.2.1 Arguments

The notion of argument is at the heart of several models developed for reasoning about defeasible information (e.g. [10, 15]), decision making (e.g. [4, 7]), practical reasoning (e.g. [5]) and modelling different types of dialogues (e.g. [3, 16]). An argument is a reason for believing a statement, choosing an option or doing an action. In most existing works on argumentation, an argument is either considered as an abstract entity whose origin and structure are not defined, or it is a logical proof of a statement where the proof is built from a knowledge base.

In our application, *image arguments* are reasons for accepting or rejecting a given image. They are built by users when rating the different tags associated with an image. The set $\mathcal{T} = \{t_1, \ldots, t_k\}$ contains all the available tags. We assume the availability of a function $\mathcal{F} : \mathcal{I} \rightarrow 2^{\mathcal{T}}$ that returns the tags associated with a given image. Note that the same tag may be associated with more than one image. A tag which is evaluated positively creates an *argument pro* the image whereas a tag which is rated negatively induces an *argument con* the image. Image arguments are also associated with a weight, which denotes the *strength* of the argument. We assume that the weight w of an image argument belongs to the finite set $\mathcal{W} = \{0, 0.1, \ldots, 0.9, 1\}$. The tuple $\langle \mathcal{I}, \mathcal{T}, \mathcal{S}, \mathcal{W} \rangle$ will be called a *theory*. An argument is defined as follows.

Definition 4.1 (Argument). Let $\langle \mathcal{I}, \mathcal{T}, \mathcal{S}, \mathcal{W} \rangle$ be a theory and $im \in \mathcal{I}$.

- An *argument pro im* is a pair $((t, v), w, im)$ where $t \in \mathcal{T}$, $v \in \mathcal{S}$ and $v > 0$.
- An *argument con im* is a pair $((t, v), w, im)$ where $t \in \mathcal{T}$, $v \in \mathcal{S}$ and $v < 0$.

The pair (t, v) is the *support* of the argument, w is its *strength* and im is its *conclusion*. The functions Tag, Val, Str and Conc return respectively the tag t of an argument $((t, v), w, im)$, its value v, its weight w and the conclusion im.

In our application, we are mainly interested in two things: i) to have a synthetic view of the opinion of a given user w.r.t. an image, and ii) to calculate whether the image can be regarded as worthy to be accepted or not. In the first case, we aggregate the image arguments of a user u_i to obtain his overall image preference r_i^*. Instead, for deciding whether an image is accepted or rejected by the whole group we define a multiple-criteria operator.

Definition 4.2 (Multiple-criteria decision). Let $\mathcal{U} = \{u_1, \ldots, u_n\}$ be a set of users, $im \in \mathcal{I}$ where $\mathcal{F}(im) = \{t_1, \ldots, t_m\}$. Given the following user's arguments:

Users/Tags	t_1	...	t_j	...	t_m	im
u_1	$(v_{1,1}, w_{1,1})$	...	$(v_{1,j}, w_{1,j})$	...	$(v_{1,m}, w_{1,m})$	r_1^*
$\vdots$	$\vdots$	$\vdots$	$\vdots$	$\vdots$	$\vdots$	$\vdots$
u_i	$(v_{i,1}, w_{i,1})$	...	$(v_{i,j}, w_{i,j})$	...	$(v_{i,m}, w_{i,m})$	r_i^*
$\vdots$	$\vdots$	$\vdots$	$\vdots$	$\vdots$	$\vdots$	$\vdots$
u_n	$(v_{n,1}, w_{n,1})$	...	$(v_{n,j}, w_{n,j})$	...	$(v_{n,m} w_{n,m},)$	r_n^*

The *overall image preference* of a user u_i denoted by $r_i^*(im)$ is defined as

$$r_i^*(im) = \frac{\sum_{1 \leq j \leq m} v_{i,j} \cdot w_{i,j}}{\sum_{1 \leq j \leq m} w_{i,j}} \qquad (4.8)$$

Then, a multiple-criteria decision operator can be defined as

$$\text{MCD}(im) = \begin{cases} 1, & \text{if } \forall u_i, 0 \leq r_i^*(im) \leq 1 \\ -1, & \text{if } \forall u_i, -1 \leq r_i^*(im) < 0 \\ 0, & \text{otherwise} \end{cases} \qquad (4.9)$$

Note that the MCD criterion allows three values: 1 (for acceptance), -1 (for rejection) and 0 (for undecidedness). Therefore, an image *im* is automatically added to the image collection if it has been unanimously accepted by the users. On the contrary, the image is discarded if it has been unanimously rejected. Finally, if $\text{MCD}(im) = 0$, then the system is unable to decide and the final decision is taken by the users in a Voting scene.

4.4.2.2 Arguing and Users' Agreement

The main goal of two users in a Forum scene is to try to reach an agreement about whether to "keep" or to "discard" an image *im* by exchanging arguments about the image. The Forum scene defines a bilateral argumentation protocol.

- Two users tag an image *im* by means of image tags, and they can propose their image tags to the other user:

  - While tagging, their overall image preferences are automatically updated;

- A user proposes an image tag to the other user who can either accept or reject it:

  - If the user accepts the image tag proposed, then their overall image preferences are automatically updated:
    - · If an *argue agreement* is reached, then the Forum scene stops,
    - · otherwise, the Forum scene keeps on;
  - if the user rejects the image tag proposed, then the Forum scene keeps on.

Both users can also decide to leave the Forum scene spontaneously.

Informally, an *argue agreement* is reached when the image preferences of the two users agree on "keep" or "discard". Let $r_i^*(im)$ and $r_j^*(im)$ be the image *im* preferences of user u_i and u_j respectively. Then, a decision criterion for deciding whether an argue agreement is reached can be defined as

$$\text{argue}(im) = \begin{cases} 1, & \text{if } (0 \leq r_i^*(im) \leq 1 \wedge 0 \leq r_j^*(im) \leq 1) \\ & \vee (-1 \leq r_i^*(im) < 0 \wedge -1 \leq r_j^*(im) < 0) \\ 0, & \text{otherwise} \end{cases} \qquad (4.10)$$

Therefore, an argue scene stops when $\texttt{argue}(im) = 1$. Instead, while $\texttt{argue}(im) = 0$, the argue scene keeps on until either $\texttt{argue}(im) = 1$ or the two users decide to stop arguing.

4.4.3 Voting Scene

The main goal of the users in a Voting scene is to decide by vote whether or not to add an image to the image collection. This decision step occurs when the automatic decision process at the end of the forum scene is unable to make a decision.

In a Voting scene, each user's vote can be "yes", "no" or "abstain" (in case that no vote is provided). Let $v_i \in \{+1, 0, -1\}$ be the vote of user u_i where $+1 =$ "yes", $-1 =$ "no" and $0 =$ "abstain" and let $\mathcal{V} = \{v_1, v_2, \ldots, v_n\}$ be the set of votes of the users in a vote scene. Then, a decision criterion for adding an image or not based on *vote counting* can be defined as

$$\texttt{vote}(im_j) = \begin{cases} 1, & \text{if } \sum_{1 \le i \le n} v_i \ge 0 \\ 0, & \text{otherwise} \end{cases} \qquad (4.11)$$

Therefore, an image im_j is added to the image collection if the number of "yes" votes is greater or equals than the number of "no" votes. In the above criterion, a neutral situation is considered as a positive vote.[1]

References

1. Amgoud, L., Confalonieri, R., de Jonge, D., d'Inverno, M., Hazelden, K., Osman, N., Prade, H., Sierra, C., Yee-King, M.: Sharing online cultural experiences: An argument-based approach. In: V. Torra, Y. Narukawa, B. López, M. Villaret (eds.) Modeling Decisions for Artificial Intelligence, *Lecture Notes in Computer Science*, vol. 7647, pp. 282–293. Springer, Berlin (2012). DOI 10.1007/978-3-642-34620-0_26. URL http://dx.doi.org/10.1007/978-3-642-34620-0_26
2. Amgoud, L., Confalonieri, R., Jonge, D.D., d'Inverno, M., Hazelden, K., Osman, N., Prade, H., Sierra, C., Yee-King, M.: WeCurate: Designing for synchronised browsing and social negotiation. In: S. Ossowski, F. Toni, G.A. Vouros (eds.) Proceedings of the First International Conference on Agreement Technologies, AT 2012, *CEUR Workshop Proceedings*, vol. 918, pp. 168–179. CEUR-WS.org (2012). URL http://ceur-ws.org/Vol-918/111110168.pdf
3. Amgoud, L., Dimopoulos, Y., Moraitis, P.: A unified and general framework for argumentation-based negotiation. In: Proc. of the 6th International Joint Conference on Autonomous Agents and Multiagent Systems (AAMAS'07), pp. 963–970. ACM Press, New York (2007)
4. Amgoud, L., Prade, H.: Using arguments for making and explaining decisions. Artificial Intelligence Journal **173**(3–4), 413–436 (2009)

---

[1] This assumption is made to avoid an *undecided* outcome in this decision step.

5. Atkinson, K., Bench-Capon, T., McBurney, P.: Justifying practical reasoning. In: F. Grasso, C. Reed, G. Carenini (eds.) Proc. of the Fourth Workshop on Computational Models of Natural Argument (CMNA'04), pp. 87–90 (2004)
6. Barron, C.: Smithsonian museum explores the art of gaming. The Washington Post (2012)
7. Bonet, B., Geffner, H.: Arguing for decisions: A qualitative model of decision making. In: Proc. of the 12th Conference on Uncertainty in Artificial Intelligence (UAI'96), pp. 98–105 (1996)
8. Confalonieri, R., Yee-King, M., Hazelden, K., d'Inverno, M., de Jonge, D., Osman, N., Sierra, C., Agmoud, L., Prade, H.: Engineering multiuser museum interactives for shared cultural experiences. Eng. Appl. Artif. Intell. 46(A), 180–195 (2015). DOI 10.1016/j.engappai.2015. 08.013. URL http://dx.doi.org/10.1016/j.engappai.2015.08.013
9. d'Inverno, M., Luck, M., Noriega, P., Rodríguez-Aguilar, J.A., Sierra, C.: Communicating open systems. Artificial Intelligence 186(0), 38–94 (2012). DOI 10.1016/j.artint. 2012.03.004. URL http://www.sciencedirect.com/science/article/pii/S0004370212000252
10. Dung, P.M.: On the acceptability of arguments and its fundamental role in nonmonotonic reasoning, logic programming and n-person games. Artificial Intelligence Journal 77(2), 321–357 (1995)
11. Gaitatzes, A., Christopoulos, D., Roussou, M.: Reviving the past: Cultural heritage meets virtual reality. In: Proceedings of the 2001 Conference on Virtual Reality, Archeology, and Cultural Heritage, VAST '01, pp. 103–110. ACM, New York (2001). DOI 10.1145/584993. 585011. URL http://doi.acm.org/10.1145/584993.585011
12. Greg, A.: Your paintings: public access and public tagging. Journal of the Scottish Society for Art History 16, 48–52 (2011)
13. Hazelden, K., Yee-King, M., Confalonieri, R., Sierra, C., Ghedini, F., de Jonge, D., Osman, N., d'Inverno, M.: WeCurate: multiuser museum interactives for shared cultural experiences. In: W.E. Mackay, S.A. Brewster, S. Bødker (eds.) 2013 ACM SIGCHI Conference on Human Factors in Computing Systems, CHI '13, pp. 571–576. ACM, New York (2013). DOI 10.1145/ 2468356.2468457. URL http://doi.acm.org/10.1145/2468356.2468457
14. Hornecker, E.: "I don't understand it either, but it is cool" - visitor interactions with a multi-touch table in a museum. In: Horizontal Interactive Human Computer Systems, 2008. TABLETOP 2008. 3rd IEEE International Workshop on, pp. 113–120. IEEE, Los Alamitos (2008). DOI 10.1109/TABLETOP.2008.4660193
15. Pollock, J.L.: How to reason defeasibly. Artificial Intelligence Journal 57, 1–42 (1992)
16. Prakken, H.: Coherence and flexibility in dialogue games for argumentation. Journal of Logic and Computation 15(6), 1009–1040 (2005)
17. Ribeiro, R.A.: Fuzzy multiple attribute decision making: A review and new preference elicitation techniques. Fuzzy Sets Syst. 78(2), 155–181 (1996). DOI 10.1016/0165-0114(95) 00166-2. URL http://dx.doi.org/10.1016/0165-0114(95)00166-2
18. Torra, V.: The weighted OWA operator. International Journal of Intelligent Systems 12(2), 153–166 (1997)
19. Turner, C.: Making native space: Cultural politics, historical narrative, and community curation at the National Museum of the American Indian. Practicing Anthropology 33(2), 40–44 (2011)
20. Wojciechowski, R., Walczak, K., White, M., Cellary, W.: Building virtual and augmented reality museum exhibitions. In: Proceedings of the Ninth International Conference on 3D Web Technology, Web3D '04, pp. 135–144. ACM, New York (2004). DOI 10.1145/985040.985060. URL http://doi.acm.org/10.1145/985040.985060
21. Yager, R.R.: On ordered weighted averaging aggregation operators in multicriteria decision-making. IEEE Trans. Syst. Man Cybern. 18(1), 183–190 (1988)
22. Yee-King, M., Confalonieri, R., de Jonge, D., Hazelden, K., Sierra, C., d'Inverno, M., Amgoud, L., Osman, N.: Multiuser museum interactives for shared cultural experiences: An agent-based approach. In: Proceedings of the 2013 International Conference on Autonomous Agents and Multi-agent Systems, AAMAS '13, pp. 917–924. International Foundation for Autonomous Agents and Multiagent Systems, Richland (2013). URL http://dl.acm.org/citation.cfm?id=2484920.2485066

23. Yee-King, M., Confalonieri, R., de Jonge, D., Osman, N., Hazelden, K., Amgoud, L., Prade, H., Sierra, C., d'Inverno, M.: Towards community browsing for shared experiences: The WeBrowse system. In: S. Ossowski, F. Toni, G.A. Vouros (eds.) Proceedings of the First International Conference on Agreement Technologies, *CEUR Workshop Proceedings*, vol. 918, pp. 201–202. CEUR-WS.org (2012). URL `http://ceur-ws.org/Vol-918/111110201.pdf`

Chapter 5
The PeerLearn Application

Dave de Jonge, Bruno Rosell, Patricia Gutierrez, Nardine Osman and Carles Sierra

There are a number of available tools that support teachers in the management of lesson plans on the web. However, none of them is task-centred or supports any form of lesson plan *execution* over the web. PeerLearn is an application that allows both the design and the execution of lesson plans, where lesson plans are designed with respect to a selected rubric. PeerLearn uses electronic institutions to coordinate interactions, ensuring the rules set by the lesson plan are followed, and it relies on a trust-based model to calculate automated marks. The automated marks provide tremendous support for teachers when their online classrooms have massive numbers of students. This chapter provides an overview of the PeerLearn application, describes its underlying electronic institution, and presents a brief introduction to its automated-assessment technology.

5.1 Objectives

Online education has regained popularity in recent years due to the MOOC phe nomenon. In this section, we present a pedagogical framework for online-learning support in the context of communities of learners. The goal is to help tutors to man-

Dave de Jonge
IIIA-CSIC, Barcelona, e-mail: `davedejonge@iiia.csic.es`

Bruno Rosell
IIIA-CSIC, Barcelona, e-mail: `rosell@iiia.csic.es`

Patricia Gutierrez
eDreams ODIGEO, Barcelona, e-mail: `patricia.gutierrez@edreamsodigeo.com`

Nardine Osman
IIIA-CSIC, Barcelona, e-mail: `nardine@iiia.csic.es`

Carles Sierra
IIIA-CSIC, Barcelona, e-mail: `sierra@iiia.csic.es`

© Springer Nature Switzerland AG 2024   97
N. Osman (ed.), *Electronic Institutions*, Artificial Intelligence: Foundations,
Theory, and Algorithms, https://doi.org/10.1007/978-3-319-65605-2_5

age (specify and monitor) pedagogical workflows that (1) easily allow interaction within groups of students and (2) include automated tools to support the task of assessing massive numbers of students, as needed in MOOCs.

In terms of management, we provide a system based on electronic institutions [3] that allows the tutor to specify pedagogical workflows. These are lesson plans that contain course steps and their interrelations (e.g. practice tasks, quiz solving, mutual assessments etc). Once the pedagogical workflow is defined, a specific graphical user interface (GUI) is automatically generated to allow students to navigate through the lesson. Every time the tutor modifies a workflow, a new GUI is generated accordingly without any programming effort. (The automatic GUI generation that is referred to in this chapter is described in detail in Chapter 6.) In terms of marking, we deal with the problem of massive numbers of students registered on online courses. Students may receive assessments from other peer students, increasing the level of interaction in the community and providing feedback about their performance. As the tutor is simply not able to mark everybody (since the number of students may reach tens of thousands), our infrastructure includes a trust-based peer-assessment service to help with the process of assessing large numbers of students. The service generates tentative marks for students by comparing peer assessments to tutor assessments. Additionally, we also make use of existing services providing automatic analysis of music performances, such as the performance-analysis service, which analyses the onsets of the student's performance with respect to the tutor's uploaded reference performance,[1] and the progress-analysis service, which compares a new performance with a previous performance and provides feedback on the progress that the student has made since his or her previous performance.

Our framework is based on electronic institutions [3], and it integrates a number of assessment and feedback services, including our trust-based peer assessments, the performance-analysis service and the progress-analysis service. We use electronic institutions for the implementation of pedagogical workflows. Tutors are able to define a learning workflow with the ISLANDER tool. This workflow is executed by the EIDE framework [8], meaning that it will provide a virtual environment for the students that automatically verifies that the students correctly upload homework assignments, and that grants them access to the next stage of the workflow only if they have successfully finished the current stage. For example, a student may not be allowed to start a level-2 course if he has not yet passed the exam of the level-1 course.

In existing online-learning systems, these kinds of rules need to be re-implemented at the code level for every new learning plan. With the EIDE framework, however, this is no longer necessary because one can define the rules in a visual manner, using ISLANDER, and the application of those rules is then automatically taken care of by the EIDE framework. As mentioned earlier, and as we illustrate in Chapter 6, the GUI which allows people to interact in an EI through a web browser can be automatically generated too, meaning that one only needs to modify the EI specification when

---

[1] In audio analysis, onset detection compares two audio files, looking for a number of differences, such as changes in spectral energy or changes in detected pitch.

needed (i.e. whenever the lesson plan needs to be modified) without having to redesign the GUI for every such modification.

The remainder of this chapter is divided as follows. Section 5.2 presents the implementation of the PeerLearn application; Section 5.3 presents the electronic institutions specification underlying the PeerLearn application; and Section 5.4 describes the integrated technologies that provide assessments and feedback of students' performances.

5.2 Illustration

In this section we present the user interface of PeerLearn [2]. We remind the reader that the user interface is automatically generated from the EI specification. As such, every time the tutor modifies her pedagogical workflow, the views of the GUI are automatically modified accordingly, and this includes the actions that one is allowed to perform at each stage of the pedagogical workflow. We also note that while PeerLearn was designed for online learning in general, the pedagogical workflow illustrated in our examples focuses on music learning.

Figures 5.1 and 5.2 present the "map" view for the tutor and student, respectively. The map view provides the user with a general idea of the pedagogical workflow and its different stages, in addition to the current stage of the user (a white box with dashed – not dotted – yellow/light grey borders) and what they need to do next (a white box with solid green/dark grey borders). Dark/Red boxes represent pedagogical steps that are not relevant to the user role viewing the map. For instance, the tutor does not need to know the details of getting the song, performing it and providing student feedback to other performances (Figure 5.1). In our specific implementation, one can see that the pedagogical workflow is specified differently for each role. The tutor first needs to upload the music sheet and audio file ("Select song" step of Figure 5.1) before moving to marking the students' performances ("Song performance evaluation" step of Figure 5.1). As for the student, he first needs to get the song uploaded by the tutor ("Get song" step of Figure 5.1), then practice and upload his own performance ("Perform" step of Figure 5.1), before assessing his fellow students' performances ("Give and receive feedback" step of Figure 5.1).

To present a sample of the interface for one of the pedagogical steps, Figure 5.3 illustrates the view where a student can upload her own performance. The actions that may be performed in this view are presented by the 'gear' icons on the left hand side, and these are: (1) Send message, where a student may send a chat message to fellow students; (2) Upload track, where a student uploads her performance; (3) Performance analysis, where an audio analysis server is requested to automatically analyse the student's performance and compare it to that of the tutor; and (4) Progress analysis, where another audio analysis server is requested to automatically analyse the student's history of performances and assess whether the student has been improving over time or not. Figure 5.4 provides an example of the performance analysis output, where the higher the line is, the further the students performance is from that of

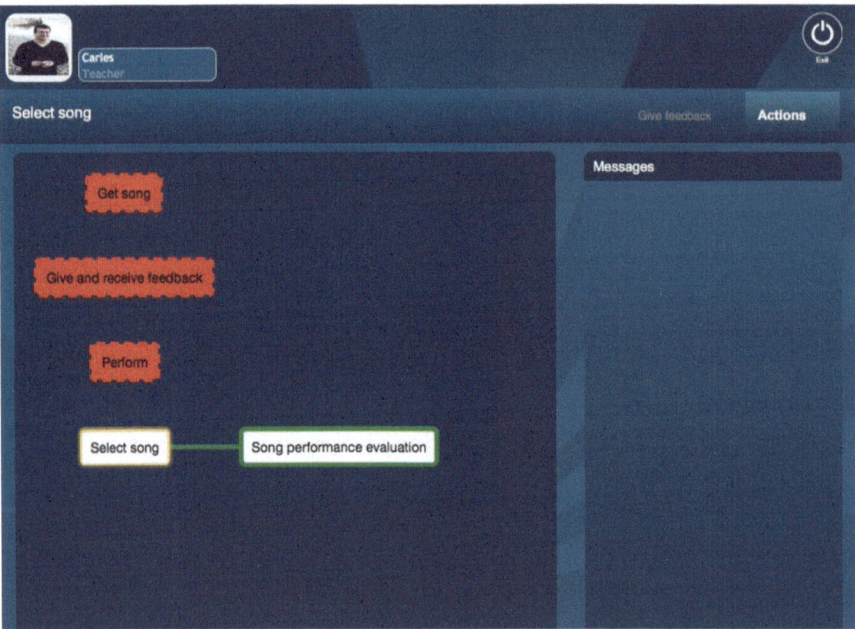

Fig. 5.1 The tutor's view of the pedagogical workflow

Fig. 5.2 The student's view of the pedagogical workflow

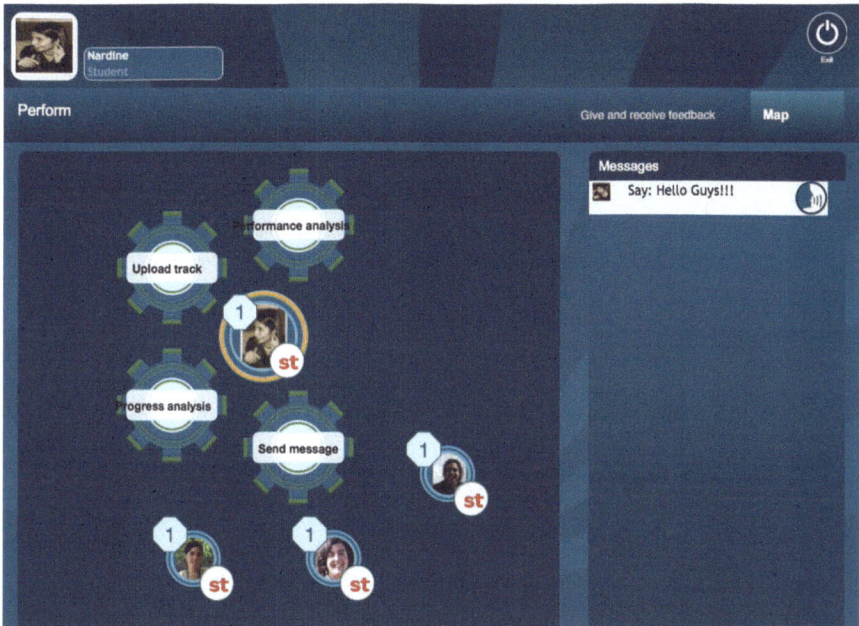

Fig. 5.3 The view for uploading one's performance

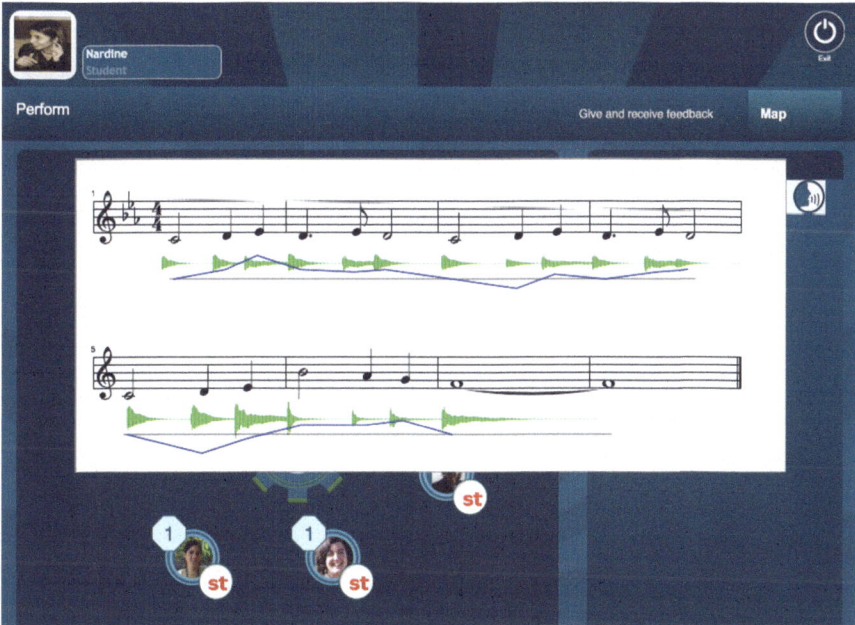

Fig. 5.4 The student's view of the pedagogical workflow

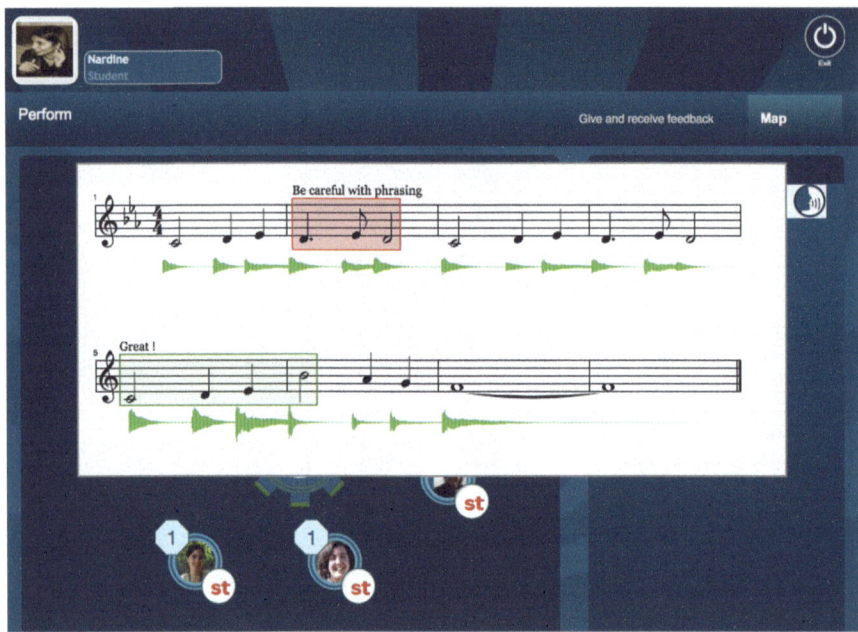

Fig. 5.5 The tutor's view of the pedagogical workflow

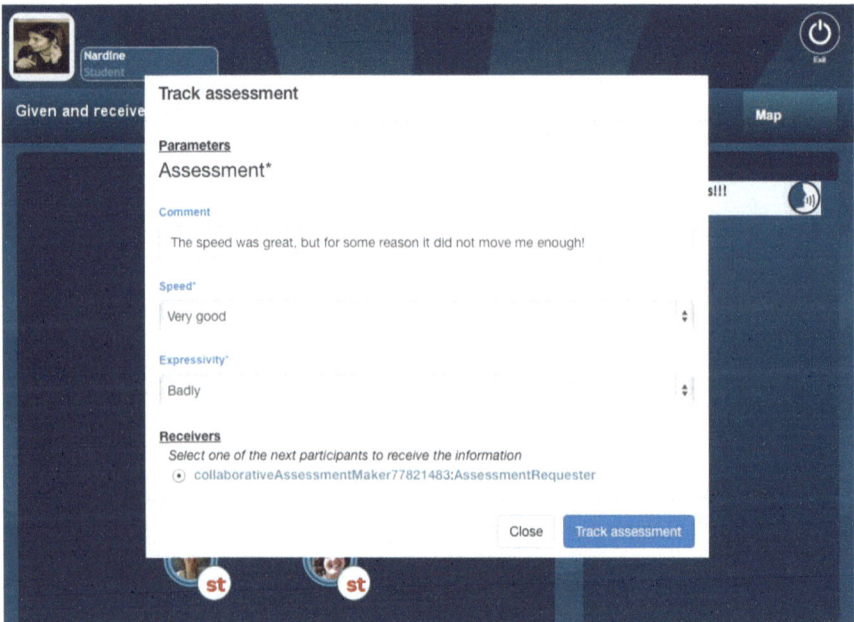

Fig. 5.6 The pop-up window for assessing one's performance

the tutor. Figure 5.5 provides an example of the progress analysis output, where green/light grey boxes indicate improvement areas and red/dark grey boxes indicate retrogression areas. Note that both performance analysis and progress analysis are services external to the EI engine, but made use of by having the EI specification call them when needed.

After performances are uploaded, the student may move to the next pedagogical step, where he may assess other students' performances. Figure 5.6 illustrates the interface for students to enter their assessments of each others' performances. Every time a new assessment is introduced, whether by a student or a tutor, the automated-assessment service (another service that the EI specification calls) recalculates the assessment of each student's performance accordingly.

5.3 The EI Specification

As an example we have implemented as an electronic institution a lesson plan that could be used for the students in the Music Circle platform, a platform where students can upload and discuss music performances.[2] The idea of this lesson plan is that there is one teacher and a group of students. The task for the students is to learn to play a certain piece of music, chosen by the teacher. The students can record their own playing and upload it to the Music Circle web page. Once it is uploaded they can apply an online tool that analyses the quality of the student's playing. Students can record, upload and analyse their music several times until they are satisfied. Then, the students share their recordings with each other and make peer assessments. Finally, the teacher also makes an assessment of a subset of the students, which will be their final mark. For the other students an automated-assessment service is called that calculates the students' final marks based on their peers' assessments. The performative structure of this institution is displayed in Figure 5.7.

The participants in the institution can play one of the following roles:

- student
- tutor

Apart from these, the institution defines two more roles that are to be played by software agents:

- database agent
- trust agent

The institution is launched by a tutor who wishes to set up a lesson. Once the institution has started the tutor enters the institution together with a database agent. The tutor and database agent both move to the 'Upload Assignment' scene, where the tutor can upload the assignment (consisting of the score file of the piece of music to study and an example audio file) to the Music Circle database. When uploading has finished the tutor sends a message to the database agent with the URL of the location

---

[2] https://goldsmiths.musiccircleproject.com

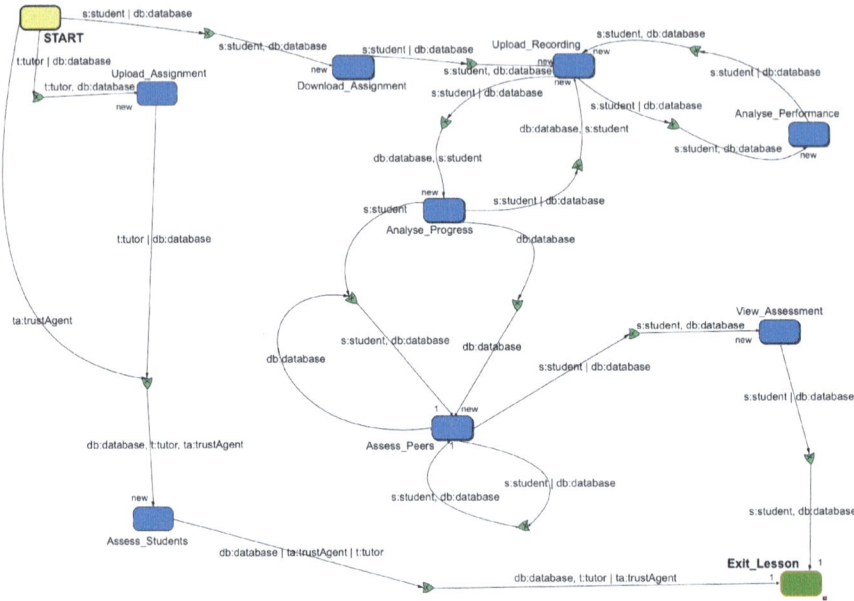

Fig. 5.7 The performative structure of an example lesson plan in Music Circle

where the file was uploaded. The database agent can then access the database to verify that the file is indeed there. In general, every database agent in the institution has free access to the database. In this way the database agent can perform the administration of the uploaded files, and provide access to certain files to the students by sending messages containing the URL of that file. When the assignment has been uploaded the tutor and database agent move on to the 'Assess Students' scene where they are accompanied by a trust agent. Here they will wait until the students have entered, and finished their assignments.

When the students enter the institution, each of them goes to a separate instance of a 'Download Assignment' scene, together with a database agent. In this scene the database agent sends a message to the student containing the URL of the files that have been uploaded by the tutor. Next, they move to the 'Upload Recording' scene. Now that the student has downloaded the score and example file he or she can start studying the piece and make a recording of him- or herself playing. Once such a recording has been made, the student can upload it in the 'Upload Recording' scene. Again, once the upload has finished successfully the student sends a message to the database agent containing the URL.

The student and database agent can now move back and forth between the 'Upload Recording' scene, the 'Analyse Progress' scene and the 'Analyse Performance' scene, as often as the student wants. The 'Analyse Performance' scene allows the student to activate a web service that performs an onset analysis of the student's recording compared with the example recording uploaded by the tutor. In the 'Analyse progress' scene the student can activate a similar web service, only this time the uploaded track is also compared with a previous track uploaded by the student, so that the web service can give feedback on the progress that the student has made since his or her previous upload. In Section 5.4.2 we explain how these web services have been integrated with the EI infrastructure. Every time the student and the database agent move from one of these three scenes to another, a new instance of the target scene is created and the instance of the scene they came from is closed.

When the student feels that he has practised enough and is satisfied with his last recording he moves, together with his database agent, to an instance of the 'Assess Peer' scene. This scene has an instance for every student that has finished practising. When the student moves to this scene, his database agent will create a new scene instance corresponding to him. However, the point of this scene is that he enters an instance corresponding to another student. For example, if the student enters the scene instance of student x, then he can download the final track of student x and give a mark to that student.

It works as follows: if student y enters the instance of student x then y sends a message to the database agent requesting to download student x's track. The database agent will then reply with a message containing the URL of that track. Student y can then make a comment on the track by sending a 'comment' message to the database agent, who will then store it in the database. Furthermore, student y can make an assessment of the track by sending an 'assess' message to the database agent, containing a mark for each of the criteria (in this case the performance speed and maturity). Again, these marks are then stored in the database by the database agent. This protocol is displayed in Figure 5.8. After marking, student y can leave the scene instance and move to another instance (corresponding to yet another student) of the 'Assess Peer' scene. A student can make as many assessments as he likes, but must make at least one assessment. Once the student feels he has made enough assessment he and his database agent move to the 'View Assessment' scene.

Meanwhile, the tutor is waiting in the 'Assess Students' scene. Whenever a student has uploaded his or her final recording, the tutor may download it and give a final mark to that student. Since there may be many students active, it may not be feasible for the tutor to mark all present students. Therefore, he is aided by a trust agent that will calculate an assessment of all the other students, based on the peer assessments. The working of the trust agent is explained in Section 5.4.1. The protocol of the 'Assess Students' scene is displayed in Figure 5.9. When all students have received a final mark, the tutor's work is done, so he can move to the 'Exit' scene and leave the institution.

When the student enters the 'View Assessment' scene, he can request his final mark. The protocol simply consists of the student sending a request message to the database agent, and the database agent responding with a message containing the final mark, if it is indeed available. If the final mark has not yet been written to the database (because the tutor or the trust agent have not yet given the mark) the database agent will respond with a 'wait' message. The student can then wait for some time and try again. When the student has received his final mark, he has completed the lesson and will therefore move to the 'Exit' scene and leave the institution.

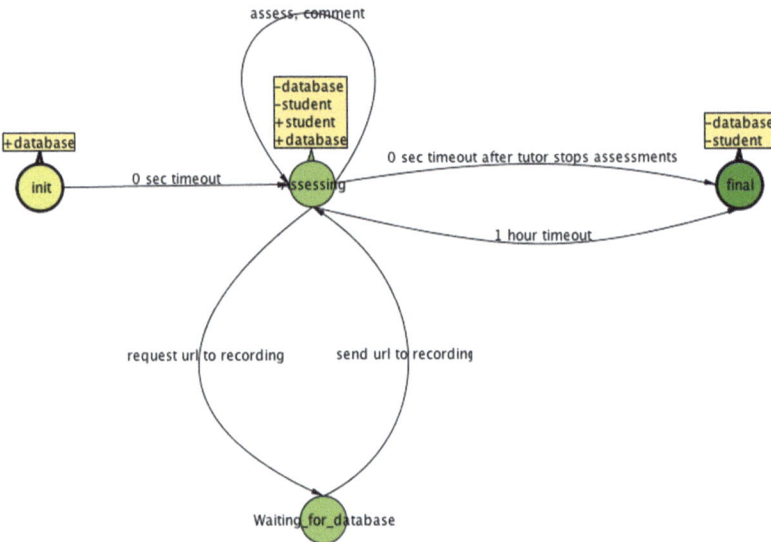

Fig. 5.8 The protocol of the Peer Assessment scene

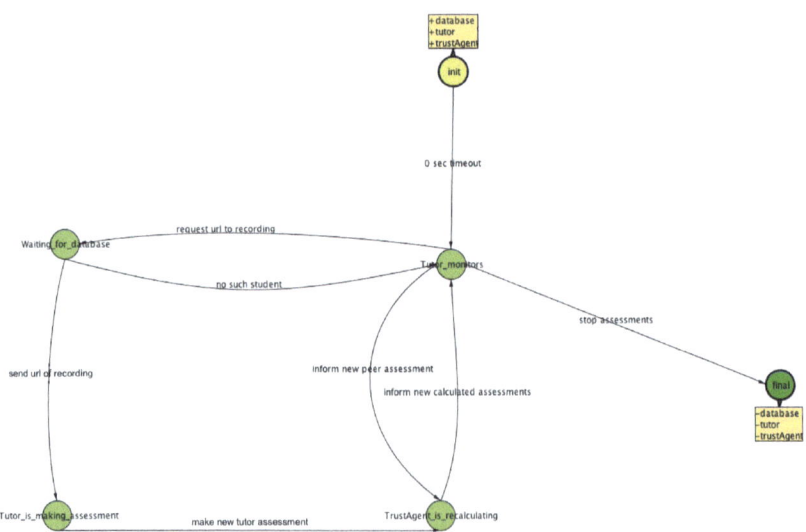

Fig. 5.9 The protocol of the Assess Students scene

5.4 Integrated Technologies

5.4.1 Collaborative Assessment

We introduce the automated-assessment service for online-learning support. This service is intended for intelligent online-learning applications that encourage student interactions, benefiting from their feedback to build trust measures and provide automated marks for students' assignments.

Self and peer assessment have clear pedagogical advantages [7, 11, 10, 5, 6]. Students increase their responsibility and autonomy, get a deeper understanding of the subject, become more active in the learning process, reflect on their role in group learning and improve their judgement skills. Online-learning communities encourage different types of peer-to-peer interactions throughout the learning process. These interactions permit students to get more feedback, to be more motivated to improve, and to compare their own work with other students' accomplishments. Tutors, on the other hand, benefit from these interactions as they get a clearer perception of the student engagement and learning process.

Peer assessment may also have the positive side effect of reducing the marking load of tutors. This is especially critical when tutors face the challenge of marking large numbers of students as needed in the increasingly popular Massive Open Online Courses (MOOC). Previous works have proposed different methods of peer assessment as part of the learning process [9, 1]. Differently from these works, we want to study the *reliability* of student assessments when compared with tutor assessments. We place more trust in students that have similar opinions to those of the tutor, and give more weight to their assessments when calculating automated marks. Although part of the learning process is that students participate in the definition of the evaluation criteria, the tutor wants to be certain that the scoring of the students' work is fair and as close as possible to his/her expert opinion.

5.4.1.1 Notation and Preliminaries

We say an online course has a tutor τ, a set of peer students S and a set of assignments $\mathcal{A}$ that need to be marked by the tutor and/or students with respect to a given set of criteria C.

The automated-assessment state S is then defined as the tuple

$$S = \langle R, \mathcal{A}, C, \mathcal{L} \rangle$$

$R = \{\tau\} \cup S$ defines the set of possible referees (or markers), where a referee could either be the tutor τ or some student $s \in S$. $\mathcal{A}$ is the set of submitted assignments that need to be marked and $C = \langle c_1, \ldots, c_n \rangle$ is the set of criteria that assignments are marked upon. $\mathcal{L}$ is the set of marks (or assessments) made by referees, such that $\mathcal{L} : R \times \mathcal{A} \rightarrow [0, \lambda]^n$ (we assume marks to be real numbers between 0 and some maximum value λ). In other words, we define a single assessment as $\mu_\alpha^\rho = \mathbf{M}$, where $\alpha \in \mathcal{A}$, $\rho \in R$ and $\mathbf{M} = \langle m_1, \ldots, m_n \rangle$ describes the marks provided by the referee on the n criteria of C, $m_i \in [0, \lambda]$.

5.4.1.2 Similarity Between Marks

We define a similarity function $sim : [0, \lambda]^n \times [0, \lambda]^n \to [0, 1]$ to determine how close two assessments μ_α^ρ and μ_α^η are. We calculate the similarity between assessments $\mu_\alpha^\rho = \{m_1, \ldots, m_n\}$ and $\mu_\alpha^\eta = \{m_1', \ldots, m_n'\}$ as follows:

$$sim(\mu_\alpha^\rho, \mu_\alpha^\eta) = 1 - \frac{\sum_{i=1}^{n} |m_i - m_i'|}{\sum_{i=1}^{n} \lambda}$$

This measure satisfies the basic properties of a fuzzy similarity [4]. Other similarity measures could be used.

5.4.1.3 Trust Relations Between Referees

Tutors need to decide how much to trust assessments made by peers. We use two different intuitions for this. First, if the tutor and the student have both assessed some common assignments, then the more similar their marks are the more trusted is the student, and vice versa. Similarly, we can define the similarity of marks of any two students by looking at the assignments evaluated by both of them. In the case that there are no assignments evaluated by the tutor and a particular student, one approach would be to simply not take that student's opinion into account because the tutor does not know how much to trust the evaluations of that student. However, in our approach, we approximate that unknown trust by looking at the chain of trust between the tutor and the student through other students. To model this we define two different types of trust relations:

- *Direct trust*: This is the trust between referees $\rho, \eta \in R$ that have at least one assignment assessed in common. The trust value is the average of similarities on the assessments over the same peers. Let the set $A_{\rho,\eta}$ be the set of all assignments that have been assessed by both referees. That is, $A_{\rho,\eta} = \{\alpha \mid \mu_\alpha^\rho \in \mathcal{L}$ and $\mu_\alpha^\eta \in \mathcal{L}\}$. Then,

$$T_D(\rho, \eta) = \frac{\sum_{\alpha \in A_{\rho,\eta}} sim(\mu_\alpha^\rho, \mu_\alpha^\eta)}{|A_{\rho,\eta}|}$$

We could also define direct trust as the conjunction of the similarities for all common assignments as

$$T_D(\rho, \eta) = \bigwedge_{\alpha \in A_{\rho,\eta}} sim(\mu_\alpha^\rho, \mu_\alpha^\eta)$$

However, this would not be practical, as a significant difference in just one assessment of those assessed by two referees would make their mutual trust very low.

- *Indirect trust*: This is the trust between referees $\rho, \eta \in R$ without any assignment assessed by both of them. We compute this trust as a transitive measure over chains of referees for which we have pairwise direct trust values. We define a trust chain as a sequence of referees $q_j = \langle \rho_1, ..., \rho_i, \rho_{i+1}, ..., \rho_{m_j} \rangle$ where $\rho_i \in R$, $\rho_1 = \rho$ and $\rho_{m_j} = \eta$ and $T_D(\rho_i, \rho_{i+1})$ is defined for all pairs (ρ_i, ρ_{i+1}) with $i \in [1, m_j - 1]$. We denote by $Q(\rho, \eta)$ the set of all trust chains between ρ and η. Thus, indirect trust is defined as the aggregation of the direct trust values over these chains as follows:

$$T_I(\rho, \eta) = \max_{q_j \in Q(\rho, \eta)} \prod_{i \in [1, m_j - 1]} T_D(\rho_i, \rho_{i+1})$$

Hence, indirect trust is based on the notion of transitivity.[3]

Ideally, we would like to not overrate the trust of a tutor on a student, that is, we would like that $T_D(a, b) \geq T_I(a, b)$ in all cases. Guaranteeing this in all cases is impossible, but we can decrease the number of over-trusted students by selecting an operator that gives low values to T_I. In particular, we prefer to use the product $\prod$ operator, because this is the t-norm that gives the smallest possible values. Other operators could be used, for instance the *min* function.

5.4.1.4 Trust Graph

To provide automated assessments, our proposed method aggregates the assessments on a given assignment taking into consideration how trusted each marker/referee is from the point of view of the tutor (i.e. taking into consideration the trust of the tutor in the referee in marking assignments). The algorithm that computes the students' final assessment is based on a graph defined as follows:

$$G = \langle R, E, w \rangle$$

where the set of nodes R is the set of referees in S, $E \subseteq R \times R$ is the set of edges between referees with direct or indirect trust relations, and $w : E \to [0, 1]$ provides the trust value. We denote by $D \subseteq E$ the set of edges that link referees with direct trust. That is, $D = \{e \in E | T_D(e) \neq \bot\}$. Similarly, $I \subset E$ denotes indirect trust,

---

[3] T_I is based on the fuzzy-based similarity relation *sim* presented before and fulfilling the $\otimes$-transitivity property: $sim(u, v) \otimes sim(v, w) \leq sim(u, w)$, $\forall u, v, w \in V$, where $\otimes$ is a t-norm [4].

$I = \{e \in E | T_I(e) \neq \perp\} \setminus D$. The w values will be used as weights to combine peer assessments and are defined as

$$w(e) = \begin{cases} T_D(e) & \text{, if } e \in D \\ T_I(e) & \text{, if } e \in I \end{cases}$$

Figure 5.10 shows examples of trust graphs with $e \in D$ (in black) and $e \in I$ (in red/light gray) for different sets of assessments $\mathcal{L}$.

5.4.1.5 Computing Collaborative Assessments

The first step is to build a trust graph from $\mathcal{L}$. Then, the final assessments are computed as follows. If the tutor marks an assignment, then the tutor mark is considered the final mark. Otherwise, a weighted average (μ_α) of the marks of student peers is calculated for this assignment, where the weight of each peer is the trust value between the tutor and that peer. Other forms of aggregation could be considered to calculate μ_α, for instance a peer assessment may be discarded if it is very far from the rest of the assessments, or if the referee's trust falls below a certain threshold.

Figure 5.10 illustrates how the trust graph evolves with new assessments. The criteria C in this example are *speed* and *maturity* and the maximum mark value is $\lambda = 10$. For simplicity we only represent referees that made assessments in $\mathcal{L}$. In Figure 5.10(a) there is one node, the tutor, who has made the first assessment over assignment ex_1. There are no links to other nodes as no one else has assessed anything yet. In (b) student Dave assesses the same exercise as the tutor and thus a link is created between them. The trust value $w(tutor, Dave) = T_D(tutor, Dave)$ is high since their marks were similar. In (c) a new assessment by Dave on ex_2 is added with no consequences on the trust graph. In (d) student Patricia adds an assessment on ex_2 that allows direct trust to be built between Dave and Patricia and indirect trust between the tutor and Patricia, through Dave. The automated assessments generated in case (d) are: $\langle 5, 5 \rangle$ for exercise 1 (which preserves the tutor's assessment) and $\langle 3.7, 3.7 \rangle$ for exercise 2 (which uses a weighted aggregation of the peers' assessments).

5.4.2 Integrating Web Services

When learning to play a musical instrument online one would typically require access to certain resources, such as example music files and score files. Also, a student should be able to record his or her performance and share it with the teacher, or fellow students. More generally, we want the users in an EI-based application to be able to access web services. The problem however is that the existing EI infrastructure only provides a mechanism for sending short text messages. It does not support the exchange of bulk data, such as audio files.

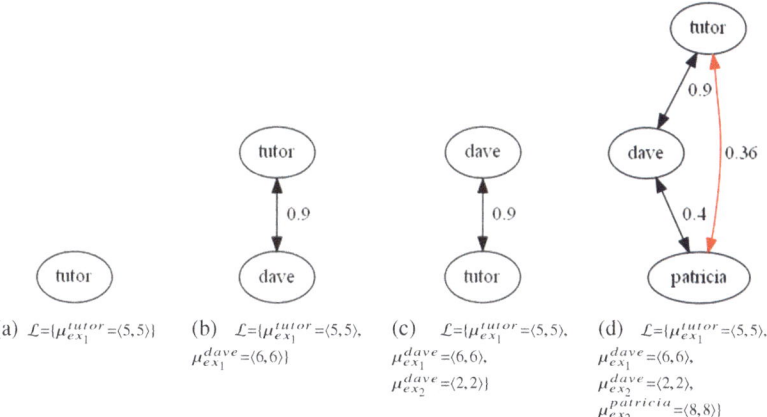

Fig. 5.10 Trust graph example

Of course, one could still allow users to exchange their files and to access web services outside the EI framework, but that would violate the purpose of the Electronic Institution. The point of using an Electronic Institution is that one can control the actions of the users, by allowing users to perform certain actions only under certain conditions. Therefore we need a mechanism that on one hand allows users to access web services and exchange files, but on the other hand makes sure that the Electronic Institution stays in control and is able to control users' access to those tools.

One can imagine for example a web service that analyses uploaded audio files and gives feedback to the user about the quality of certain technical aspects. Such a tool would require a lot of CPU time on the server and therefore we may want to restrict users to upload a file to this service, say, once a day. Also we could provide more access to the service to premium users.

We describe here how we have solved this problem with the existing Electronic Institutions framework. As an example we take a web service in which a user sends an audio file, containing music played by the user with a musical instrument, together with a file that represents the score of the music, to a web service which then analyses the *'onset'* of the musical notes. The results of this analysis are then sent back to the user in *HTML, JSON* or *XML* format. Using this tool works as follows (see also Figure 5.11):

1. The user clicks 'open webservice', which causes an http request to be sent to the EI-server, which passes the request on to the *GuiAgent*.
2. The *GuiAgent* sends an EI message within the institution to a *WebServiceAgent*.
3. The WebServiceAgent sends an *'access granted'* or *'access denied'* message, with the URL of the web service, back to the *GuiAgent*.
4. The *GuiAgent* passes this message back to the browser in an http response.
5. The browser then opens the URL in a separate window.

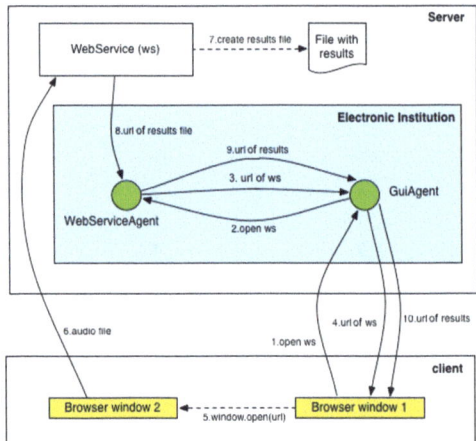

Fig. 5.11 Interaction between a user and a web service, mediated by an EI. Solid arrows represent exchange of messages, dashed arrows represent the creation of a resource. The arrows are numbered in chronological order

6. The new window enables the user to send data to the web service. For example, it allows the user to choose an audio file to upload. The user clicks 'submit' to send the data. This will cause the data to indeed be sent to the web service, but also sends a notification to the EI server.
7. When the web service has analysed the audio, the results are stored on the server.
8. The web service notifies the *WebServiceAgent* that the data is ready.
9. The *WebServiceAgent* sends an EI Message to the *GuiAgent* including the URL of the file with the results.
10. The *GuiAgent* sends the URL to the browser, which can then use this URL to load the data.

Note that indeed the EI is in control of the interactions between the user and the web service. The user cannot access the service without permission from the EI, since the user needs the URL of the web service. The EI could send, together with the URL, a password that can be used only once, to make sure that the user needs to request permission from the EI every time.

We have assumed in the above that the web service provides its own GUI, which needs to be opened in a separate browser window. Of course, this is not necessary, and the GUI to the web service may as well be fully integrated with the EI GUI. In that case step 5 can be skipped.

References

1. de Alfaro, L., Shavlovsky, M.: CrowdGrader: Crowdsourcing the evaluation of homework assignments. Technical Report 1308.5273, arXiv.org (2013)
2. Brito, I., Gutierrez, P., Hazelden, K., de Jonge, D., Lemus, L., Osman, N., Rosell, B., Sierra, C., Roig, C.: Collaborative peer assessment using PeerLearn. In: L. Steels (ed.) Music Learning with Massive Open Online Courses (MOOCs), *The Future of Learning*, vol. 6, chap. 10, pp. 145–174. IOS Press, Amsterdam (2015). DOI 10.3233/978-1-61499-593-7-145. URL http://dx.doi.org/10.3233/978-1-61499-593-7-145
3. d'Inverno, M., Luck, M., Noriega, P., Rodríguez-Aguilar, J.A., Sierra, C.: Communicating open systems. Artificial Intelligence **186**(0), 38–94 (2012). DOI 10.1016/j.artint. 2012.03.004. URL http://www.sciencedirect.com/science/article/pii/ S0004370212000252
4. Godo, L., Rodríguez, R.O.: Logical approaches to fuzzy similarity-based reasoning: an overview. In: G. Della Riccia, D. Dubois, R. Kruse, H.J. Lenz (eds.) Preferences and Similarities, pp. 75–128. Springer, Vienna (2008)
5. Hannon, V.: 'Only connect!' : a new paradigm for learning innovation in the 21st century. Centre for Strategic Education occasional paper ; no. 112, September 2009. Centre for Strategic Education, East Melbourne (2009)
6. Jenkins, H.: Confronting the Challenges of Participatory Culture: Media Education for the 21st Century. MIT Press, Cambridge (2009)
7. Lu, J., Zhang, Z.: Understanding the effectiveness of online peer assessment: A path model. Journal of Educational Computing Research **46**(3), 313–333 (2012). DOI 10.2190/EC.46.3.f
8. Noriega, P., de Jonge, D.: Electronic institutions: The EI/EIDE framework. In: H. Aldewereld, O. Boissier, V. Dignum, P. Noriega, J. Padget (eds.) Social Coordination Frameworks for Social Technical Systems, pp. 47–76. Springer International Publishing, Cham (2016). DOI 10.1007/978-3-319-33570-4_4. URL http://dx.doi.org/10.1007/ 978-3-319-33570-4_4
9. Piech, C., Huang, J., Chen, Z., Do, C., Ng, A., Koller, D.: Tuned models of peer assessment in MOOCs. In: Proc. of the 6th International Conference on Educational Data Mining (EDM 2013). International Educational Data Mining Society (2013)
10. Stepanyan, K., Mather, R., Jones, H., Lusuardi, C.: Student engagement with peer assessment: A review of pedagogical design and technologies. In: M. Spaniol, Q. Li, R. Klamma, R.W.H. Lau (eds.) Advances in Web Based LearningâÄŽ ICWL 2009, *Lecture Notes in Computer Science*, vol. 5686, pp. 367–375. Springer, Berlin (2009). DOI 10.1007/978-3-642-03426-8_44
11. Topping, K.: Peer assessment between students in colleges and universities. Review of Educational Research **68**(3), 249–276 (1998). DOI 10.3102/00346543068003249

Part III
Peer-to-Peer Electronic Institutions

Chapter 6
PeerFlow: Peer-to-Peer Electronic Institutions

Dave de Jonge, Bruno Rosell and Carles Sierra

Our aim is for electronic institutions to become a pervasive mechanism to co-ordinate very large networks of humans and devices, and thus a centralised approach carries numerous challenges for the future. Peer-to-peer (P2P) networks appear to be a natural option to implement distributed electronic institutions as they provide a number of desirable properties: they are robust, they scale well and nodes are equally privileged. In the last decade many applications and technologies have been built using P2P networks, such as file sharing, distributed storage or even web search engines. This chapter proposes a P2P implementation of electronic institutions called PeerFlow and a GUI that allows for the automatic generation of user interfaces over web browsers.

6.1 Objectives

In a recent document by IBM, "Device Democracy: Saving the future of the Internet of Things"[1] a case is made about the uncertain future of centralised approaches in the context of networks composed of billions of interconnected devises. Centralised approaches would become prohibitively expensive, would not protect privacy and would not make business models endure.

It is our ambition that electronic institutions become a pervasive mechanism to coordinate very large networks of humans and devices, and thus a centralised approach seems a bad solution for the future.

Dave de Jonge
IIIA-CSIC, Barcelona, e-mail: davedejonge@iiia.csic.es

Bruno Rosell
IIIA-CSIC, Barcelona, e-mail: rosell@iiia.csic.es

Carles Sierra
IIIA-CSIC, Barcelona, e-mail: sierra@iiia.csic.es

[1] ibm.biz/devicedemocracy

© Springer Nature Switzerland AG 2024
N. Osman (ed.), *Electronic Institutions*, Artificial Intelligence: Foundations, Theory, and Algorithms, https://doi.org/10.1007/978-3-319-65605-2_6

Peer-to-peer (P2P) networks appear to be a natural option to implement distributed electronic institutions as they provide a number of desirable properties. They are very robust in that there is no single point of failure. If a node fails, the other nodes may continue the computation, in many cases without any loss or with minimal loss of information. P2P networks scale very well as a new node increases the overall number of computation requests but also brings in resources in the form of e.g. memory or CPU cycles. Nodes are equally privileged and the quality of service they receive does not depend on whether they can afford expensive cloud servers.

In the last decade many applications and technologies have been built using this approach: file sharing (e.g. Gnutella or BitTorrent), distributed storage (e.g. Symform, Freenet, Yvi), or web search engines (e.g. Yaci, Faroo). P2P platforms (i.e. sets of interconnected nodes over TCP/IP protocols) provide different features: trust, authentication, data persistency, state guarantee, anonymity etc.

In Section 6.1.1 we describe a P2P implementation of electronic institutions called PeerFlow and in Section 6.2 a tool called Genuine that allows for the automatic generation of user interfaces over web browsers.

6.1.1 Architecture

The PeerFlow architecture is an evolution of the previous environment for running electronic institutions named Agent Mediated for Electronic Institutions (AMELI) [3]. The main features of AMELI are:

- To provide a way for different agents with different architectures to communicate with one another without any assumption about their respective internal architectures.
- To enforce upon the agents a protocol of behaviour as specified in an institution specification. This means that AMELI makes sure that the agents can only do those actions that the protocol allows them to do.

Figure 6.1 describes the different parts of AMELI. It consists of three layers: a communication layer, which enables the agents to exchange messages, a layer consisting of the agents that are acting in the institution, and in between a social layer, which controls the behaviour of the participating agents. The social layer consists of the following agents:

Governor.   For each participating agent in an institution there is one *Governor*. This *Governor* forms the link between a participating agent and the Electronic Institution. Every time the participating agent wants to perform an action in the institution, this is communicated to the *Governor* in the form of a message. The *Governor* determines whether the participating agent is indeed allowed to do that action at that moment and, if so, forwards the message to other agents in the social layer in order for the action to have effect. If the participating agent however is not allowed to do that action, the *Governor* simply blocks the message and hence

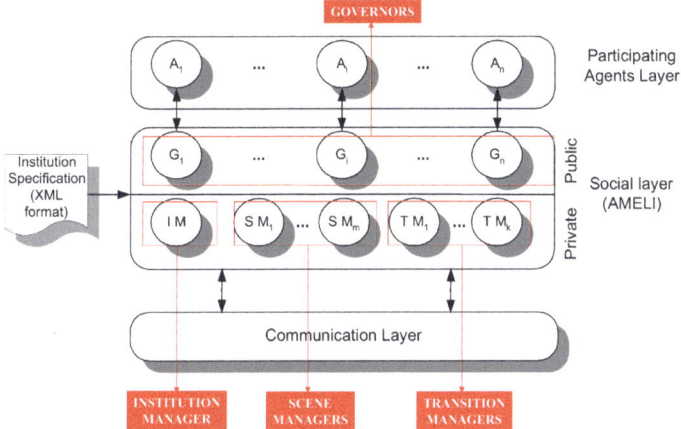

Fig. 6.1 Scheme of the social layer

prevents the action from being executed. Furthermore, the *Governor* keeps the status of the agent in the Electronic Institution and its properties.

SceneManagers. There is a scene manager for each instance of a scene protocol that is active in the electronic institution. This agent controls and keeps the status of a scene protocol of an Electronic Institution. So, it controls when an agent can enter or exit the scene protocol, the scene state, and when an agent can say a message to another.

TransitionManagers. There is one transition manager for each transition in each active performative structure. This agent controls and synchronises the movements of the agents between the nodes in a performative structure.

EInstitutionManager. This agent is unique to each running instance of an Electronic Institution and it keeps the status of the Electronic Institution. It also allows or denies an agent entry into the Electronic Institution.

The AMELI approach does not allow us to run dynamic Electronic Institutions and this is the main reason we have evolved it and create PeerFlow. It is a new AMELI that runs in a peer-to-peer environment. This provides us with some advantages, as for example a peer-to-peer network is able to maintain itself. That is, we do not need to maintain a server. This is a big advantage as it implies we do not need anybody to take care of the server maintenance. The platform can be regulated by the user community itself.

Another advantage is that a peer-to-peer network automatically scales with its number of users. Especially when the users are sending large data files and streams such as audio and video over the network, it is very beneficial if we do not have to rely on a server that may not be able to handle large amounts of data.

Furthermore, the use of a peer-to-peer network allows users to store their files locally, rather than on a server. This can be an advantage for privacy reasons as people may want to stay in control of their files.

Finally, the dynamic structure of peer-to-peer networks could help us to create dynamic communities without the necessity to have a central server to coordinate all the users. Currently, the AMELI distribution platform is launched using a configuration file in XML format that contains the information necessary for the Electronic Institution to run and the addresses of the nodes that form the peer-to-peer network. So, if any of this information needs to be modified one has to close the platform, edit the configuration file with the required changes, and launch the platform again. The existing framework therefore does not allow for the dynamic infrastructure that is required by the project.

These new features add to AMELI the *DeviceManager*, which manages the peer-to-peer environment, and the *RESTGovernor*, which allows any user to interact using different devices.

6.1.2 DeviceManager

A premature P2P version of AMELI already existed before PeerFlow, but this only allowed the creation of a P2P EI based on a configuration file that stored the information about the machines in a network. This file had to be created before the instantiation of the institution and hence it was impossible to add new nodes to the network at runtime.

Figure 6.2 displays a schematic overview of the P2P EI infrastructure. It is distributed over several hosts, which are connected by a peer-to-peer network. Each host with a P2P connection is called a *node* of the network, and is managed by an agent called the *DeviceManager*. The main tasks of this agent are:

- To provide and manage a repository of Electronic Institution specifications.
- To maintain the peer-to-peer network.
- To execute Electronic Institutions.

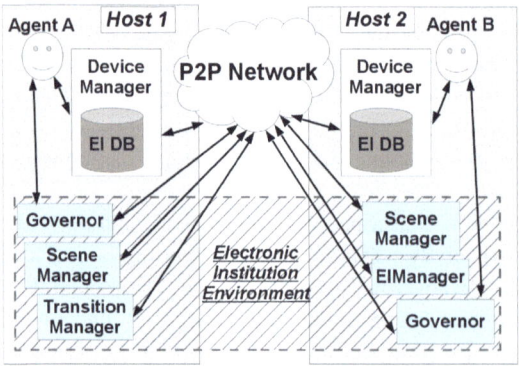

Fig. 6.2 Diagram of peer-to-peer Electronic Institutions components

6.1.2.1 Electronic Institutions Repository

One of the necessary components to run an Electronic Institution is the specification that describes the rules that the agents have to follow. The *DeviceManager* provides a repository of Electronic Institution specifications that can be used to launch a new instance of an Electronic Institution. This repository is managed by a content management system library called *Apache Jackrabbit*.[2] This allows us to store specifications in a tree structure, and also associate some properties with them.

Users or agents can interact with the *DeviceManager* to add a new specification to the repository, remove a specification from the repository or perform a search through all the specifications that are in the repository.

6.1.2.2 Peer-to-Peer Network

To simplify the management of the peer-to-peer network that the *DeviceManager* uses for communication with other *DeviceManager*s, or to launch an instance of an Electronic Institution (see Figure 6.2), we have decided to use the the *FreePastry* library.[3] This library creates a *Pastry* peer-to-peer network [2, 1, 4], and provides some useful features such as the routing of messages, or the possibility to create broadcast messages.

On the other hand, *FreePastry* does not have any feature for the discovery of new nodes that can be added to the peer-to-peer network. Therefore, we have created a simple discovery mechanism using *UDP* packet broadcasting. This process begins when a new *DeviceManager* is launched. This agent then opens a socket to listen for *UDP* requests and sends a broadcast message over the local network with the P2P node information. When other *DeviceManager*s receive the broadcast they use the received information to add the node information to their P2P table and reply with their information. After that the original *Device Manager* uses the received information to update its internal P2P table.

6.1.2.3 Executing Electronic Institutions

Another feature of the *DeviceManager* is the creation of Electronic Institution instances dynamically. That is, it is capable of creating a running AMELI infrastructure where the agents can interact following the protocols defined in a given Electronic Institution specification without the necessity of creating a configuration file. The creation of this Electronic Institution is divided into two stages: in the first stage the agents that form the social layer are created, while in the second stage the *Governors* of the participating agents are created.

---

[2] http://jackrabbit.apache.org/

[3] http://www.freepastry.org/FreePastry/

The launch of a new Electronic Institution instance starts with an agent requesting a *DeviceManager* to create the infrastructure with a given EI specification from the repository. Next, the *DeviceManager* and the other *DeviceManager*s that are in the P2P network (see Figure 6.2) cooperatively launch all the components necessary for a running Electronic Institution instance. This allows the distribution of workload between the hosts that form the P2P network. Thus, the agents that form part of the social layer (EIManager, SceneManager etc.) are created in the different nodes following a round-robin algorithm and they use the same P2P network created by the *DeviceManager*s to communicate with one another. As soon as all the necessary components have been launched, the agent is informed that the institution is ready. The next step for the agent is to request the *DeviceManager* to create a *Governor*. Once this *Governor* is created it opens a socket server where it waits for requests from the agent to send messages in the Electronic Institution. It also uses the same socket to send the agent information on the Electronic Institution state or messages that it receives from other agents that are participating in the same scenes.

Instead of launching a new instance of an Electronic Institution, one can also join one that is already running. In this case the agent would ask the *DeviceManager* for a list of all running instances that execute a given EI specification. The *DeviceManager* then uses the P2P network to send a broadcast message and receive information about the active instances in any part of the distributed system that are executing the given specification. The *DeviceManager* will then send a list of those running instances back to the agent. After that the agent can request that a *Governor* be created so that it can participate in one of them.

When a *Governor* has been created for a participating agent, this *Governor* can only be killed after the agent has left the Electronic Institution.

6.1.3 RESTGovernor

The *RESTGovernor* can manage interactions with humans in an institution using different devices such as tablets, computers or mobile phones. It is a combination of a *Jetty*[4] web server and an agent that can participate in a Peer-To-Peer Electronic Institution.

As can be seen in Figure 6.3, whenever a user does some kind of action on his or her device (such as a tablet, phone or laptop) the device sends an *HTTP* request to the *RESTGovernor*, which then sends the request to the proper agent in the institution. For example, if the user requests to search for a published specification, an HTTP request containing the search query is sent to the *RESTGovernor*. The *RESTGovernor* forwards this request to the *DeviceManager* (because the *DeviceManager* is responsible for managing the database of published specifications), and when the *RESTGovernor* receives the response it sends the result back to the user in the form of a *JSON* object.

---

[4] http://www.eclipse.org/jetty/

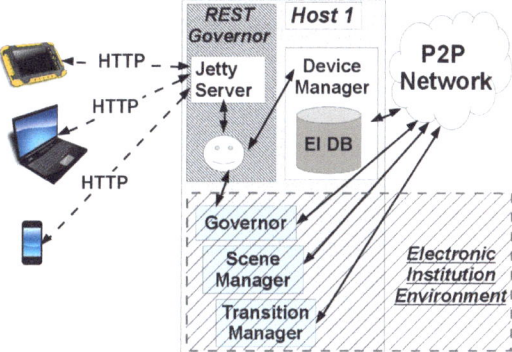

Fig. 6.3 Diagram of the interaction of the REST Governor with the peer-to-peer Electronic Institutions components

The actions that a user can do through the *RestGovernor* are:

Registering a Governor.   This is used to log in and create a *RESTGovernor* for a user.

Logging out.   This terminates the *RESTGovernor* of a user.

Searching for an Electronic Institution.   With this action, a user can obtain the published Electronic Institutions that satisfy a given search query.

Registering an Electronic Institution specification.   This is used by the user to publish an Electronic Institution specification, which can later be searched for and used by other users.

Getting information about a registered Electronic Institution specification.   A user can use this action to obtain the name, description, keywords and input roles of a published Electronic Institution.

Searching for all running Electronic Institutions.   With this action a user can obtain the identifiers of the running Electronic Institutions.

Launching a new Electronic Institution.   A user can use this action to launch a new Electronic Institution that instantiates a given EI specification.

Enter to participate in an Electronic Institution.   With this action a user can join a running Electronic Institution and interact with other agents following the rules of a published EI specification.

Obtaining information about all Electronic Institutions one is participating in.   This action returns the identifiers of all the Electronic Institutions in which the user is participating.

Obtaining information about an Electronic Institution conversation.   This action can be used to obtain information about the state of a conversation in which the agent is participating.

Getting information about possible movements.   With this action, a user can obtain the possible ways it can move from one scene to another.

Moving to another conversation.   With this action an agent specifies to which scene instance it wants to move.

Getting information about the possible messages to say.   This action returns the patterns of the messages that the user can send in the current state of a scene.

Saying a message in a scene.   This is used by an agent to send a message inside a scene.

Exiting a scene.   This is used to request to exit a scene in order to move to another scene.

Do a 'Stay And Go' in a scene.   This is used by an agent to go to another scene, while staying in the current scene at the same time.

6.2 Automated GUI

Originally, the Electronic Institution framework was designed with the goal of enabling structured communication between software agents. However, many practical applications are not purely software driven, but require human input. Several examples of such applications have been presented in earlier chapters of this book. In fact, those examples were mainly centred around human users aided by software agents to automate certain tasks.

The necessity to allow humans to interact in Electronic Institutions means that we need to provide a user interface so that users can indeed connect to and interact within an Electronic Institution through a web browser. Therefore, we have implemented a generic user interface that is automatically generated based on the institution specification. This component is called GENUINE, which stands for GENerated User INterface for Electronic institutions. It is integrated with PeerFlow, but it can also be used outside PeerFlow, in a server-client-based architecture.

6.2.1 GENUINE

A human user interacts in an EI by clicking buttons in a browser window. To allow these actions to have effect in the EI, we have implemented a software agent that can represent the user inside the EI and that executes the actions requested by the user. This agent is called the *GuiAgent*. Although the current implementation does not do anything autonomously, if necessary it can be extended to give it more sophisticated capabilities, such as giving intelligent strategic advice to the user. Figure 6.4 visually illustrates how human users can now interact in an EI, as opposed to a classic EI with only software agents.

Apart from the *GuiAgent* we need to have a GUI in the form of a website that allows the user to interact. On one hand, one may want to have a good-looking GUI that is specifically designed for a given institution. But on the other hand, one may not want to develop an entirely new GUI for every new institution, or one may want to have a generic GUI available to test a new EI during its development, so that one can postpone the design of its final GUI until the EI is finished and tested. Therefore,

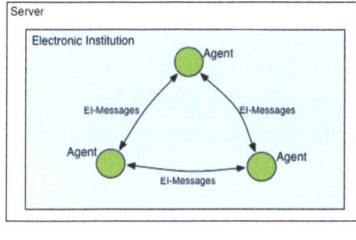
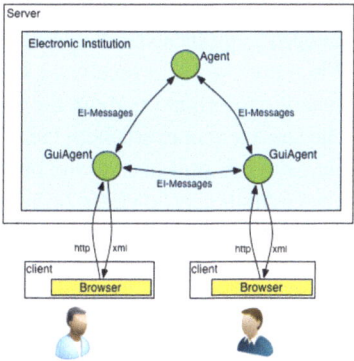

(a) Classic EI with only software agents

(b) An EI with one software agent and two human users participating

Fig. 6.4 Interaction with an electronic institution, with and without human users

we have built a framework that allows for both. It generates a GUI automatically from the EI specification, but at the same time provides an API so that web designers can easily create a custom GUI for each new EI. The main components are:

- A Java agent called *GuiAgent* that acts in the Institution on behalf of the human user.
- A Java component that encodes all relevant information the agent has about the current state of the institution into an XML file.
- A Javascript library called *GenuineConnection* that translates the XML file into a Javascript object called *EiStateInfo* that represents the information about the Electronic Institution.
- A Javascript library called *GenuinePeerflowGUI* that generates a default Graphic User Interface (as HTML) based on the *EiStateInfo* object.

6.2.2 How It Works

Whenever a user joins a running institution in the PeerFlow framework, a *GuiAgent* is automatically launched on one of the nodes in the P2P network and the *GenuineConnection*.

The process then continues as follows:

1. A web page including both Javascript libraries is loaded into the browser of the user.
2. The *GenuineConnection* library establishes a connection to this *GuiAgent* through AJAX.
3. When the *GuiAgent* is instantiated it analyses the EI specification to retrieve all static information about the institution (see below).

4. The page starts a polling service that periodically (typically several times per second) requests a status update from the *GuiAgent*.
5. When the *GuiAgent* receives a status update request it asks its *Governor* for the dynamic information about the current status of the institution.
6. The *GuiAgent* converts both the static and the dynamic information into an XML file, which is sent back to the browser.
7. The *GenuinePeerflowGUI* Javascript library then uses this information to update the user interface (more information about this below).
8. The user can now execute actions in the institution or move between its scenes by clicking buttons on the web page.
9. For each action the user makes, an HTTP request is sent to the *GuiAgent*.
10. The *GuiAgent* uses the information from the HTTP request to create an EI message, which is sent like any other message in a standard EI.

As explained, the *GuiAgent* uses two sources of information: static information from the EI specification stored on the hard disk of the server and dynamic information from the *Governor*. The static information consists of:

- The names and protocols of the scenes defined in the institution.
- The roles defined in the institution.
- The ontology of the institution.

The dynamic information consists of:

- The actions the user can do in the current scene.
- For each of these actions: the parameters to be filled out by the user.
- Which agents are present in the current scene.
- Whether it is allowed to leave the scene and, if yes, to which other scenes the user can move.

6.2.3 Generating the GUI

Every time the browser receives information from the *GuiAgent*, it updates the GUI, which takes place in two steps, handled by the two respective Javascript libraries (Figure 6.5). In the first step the *GenuineConnection* converts the received XML into a Javascript object called *EiStateInfo*, which is composed of smaller objects that represent the static and dynamic information as explained above.

In the second step the *EiStateInfo*-object is used by the *GenuinePeerflowGUI* library to draw the GUI. This GUI is completely generic, so it looks the same for every institution. If one wants to design a more fancy user interface tailored to one specific institution, one can write a new library that replaces the *GenuinePeerflowGUI*.

The fact that these two steps are handled by two different libraries enables a designer to reuse the *GenuineConnection* when designing a new GUI, so he does not have to worry about how to retrieve the relevant information from the EI. All

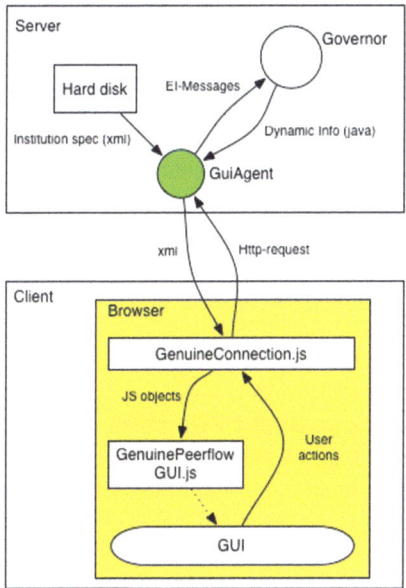

Fig. 6.5 The components necessary to generate the GUI. Solid arrows indicate exchange of information. The dashed arrow indicates that the GUI is created by the *GenuinePeerflowGUI*. Note that the machine labelled as 'Server' is in fact a node in the peer-to-peer network

information will be readily available in the *EiStateInfo*-object, so you only need to worry about how to display it on the screen.

6.2.4 The Default User Interface

An example user interface is displayed in Figure 5.3. It is divided into five sections:

- A section at the top with a profile image of the user, and his or her user name and role, along with an 'exit' button.
- A top menu, where the left part shows the name of the current scene or activity and its description, and the right part shows links to the other scenes (allowing one to navigate from scene to scene), along with a link to the map of all scenes.
- A panel to the left, showing all participants that are currently in the scene together with the user, as well as all actions that can be performed by the user.
- A panel to the right, displaying all the messages sent by/to the user (including messages from/to agents).

The panel on the left is where the user interacts with the other users and agents in the EI. The user can choose which action to perform by selecting one of the gears.

This panel only shows those actions that the user can currently do, hence preventing the user from sending illegal messages. (Note that even if the user were able to send illegal messages, they would still be blocked by the *Governor*. However, for the sake of user-friendliness we only want to display messages that the user can indeed send.) The available actions depend on the scene and the institution. If the user moves to a different scene he or she will have different actions available. These actions are defined in the institution specification.

Once the user has selected an action to take, the browser will display a form that enables the user to fill out the necessary parameters. Since each action in the institution is modelled as a message, the user needs to choose an agent that is going to receive the message. The EI specification defines which agents can receive which kinds of message. A bid in an auction house for example should always be sent to the agent playing the role of auctioneer.

The form shows one input control for each parameter of the message. The type of control depends on the type of the parameter. For example, if the parameter is of type integer, a numeric input control appears, while if the parameter is of type string, a text box appears. In case the parameter is of a user-defined type, a sub-form appears with several controls, one for each of the variables of the user-defined type. Figure 5.6 shows one example.

6.2.5 Customizing the GUI

In order to draw a GUI, a custom GUI generator can use the information from the *EiStateInfo*-object. For example: the user chooses to make a bid in an auction. The GUI generator reads from the *EiStateInfo*-object that an integer parameter must be set to represent the amount of money the user wants to bid. A GUI designer should make sure that whenever a parameter of type integer is required the GUI displays an input control that allows the user to introduce an integer value.

The fact that one can also define user-defined types in an EI allows for a lot of flexibility. Suppose for example that one would like a user to record an audio file and send this in a message to another agent. Electronic Institutions do not support audio files by default. However, the institution designer may introduce a user-defined type with the name 'Audio'. Once the user chooses to send a message that includes audio, the *EiStateInfo*-object will indicate that a parameter of type Audio is needed. A customised GUI generator can then be programmed such that for example a microphone is activated whenever this type of parameter is required.

References

1. Castro, M., Druschel, P., Kermarrec, A.M., Rowstron, A.: One ring to rule them all: service discovery and binding in structured peer-to-peer overlay networks. In: Proceedings of the 10th

ACM SIGOPS European Workshop, pp. 140–145. ACM, New York (2002)

2. Draschel, P., Rowstron, A.: Pastry: Scalable, distributed object location and routing for large-scale peer-to-peer systems. In: Proceedings of the 18th IFIP/ACM International Conference on Distributed Systems Platforms (Middleware 2001), pp. 329–350. ACM, New York (2001)

3. Esteva, M., Rosell, B., Rodríguez-Aguilar, J.A., Arcos, J.L.: AMELI: An agent-based middleware for electronic institutions. In: Proceedings of the Third International Joint Conference on Autonomous Agents and Multiagent Systems - Volume 1, AAMAS '04, pp. 236–243. IEEE Computer Society, Washington, DC (2004). DOI 10.1109/AAMAS.2004.56

4. Rowstron, A., Kermarrec, A.M., Castro, M., Druschel, P.: SCRIBE: The design of a large-scale event notification infrastructure. In: J. Crowcroft, M. Hofmann (eds.) Networked Group Communication, *Lecture Notes in Computer Science*, vol. 2233, pp. 30–43. Springer, Berlin (2001)

GPSR Compliance

The European Union's (EU) General Product Safety Regulation (GPSR) is a set of rules that requires consumer products to be safe and our obligations to ensure this.

If you have any concerns about our products, you can contact us on ProductSafety@springernature.com

In case Publisher is established outside the EU, the EU authorized representative is:

Springer Nature Customer Service Center GmbH
Europaplatz 3
69115 Heidelberg, Germany

The manufacturer's authorised representative in the EU is Springer
Nature Customer Service Centre GmbH, Europaplatz 3, 69115 Heidelberg,
Germany. If you have any concerns regarding our products, please
contact ProductSafety@springernature.com

Printed and bound by CPI Group (UK) Ltd, Croydon, CR0 4YY
29/04/2026
02099455-0016